SubjectMath.com Practice Test #1

A Full Practice Test For the Subject Math Exam
www.SubjectMath.com

2020 Edition

Copyright © 2015-2020 by Gilad Pagi, GP Group.

All rights reserved. No part of this publication may be reproduced, distributed, or transmitted in any form or by any means, including photocopying, recording, or other electronic or mechanical methods, without the prior written permission of the publisher. For permission requests, email to the publisher at the address below: GP Group,

info@gpgroupcompany.com

Printed by CreateSpace, an Amazon.com Company.

Preface

Testimonials

By Prof. Karen E Smith (Associate Chair for Graduate Students, Math Department, University of Michigan):

The website SubjectMath.com offers a valuable resource for students planning to apply to graduate programs in mathematics and the mathematical sciences in the United States. The GRE® subject examination in mathematics is required by most departments for admission to graduate programs. The study course provided at SubjectMath.com offers clear, concise online lectures on all the topics tested on the exam in a well-organized series of modules. Students can complete just the modules they most need to review, or, given that the fast pace of the exam is often the main difficulty in achieving a high score, students can complete all the modules, where they will learn many useful tricks for getting through the problems *quickly* and accurately. SubjectMath.com also provides a series of practice test booklets that are formatted exactly as the actual test, allowing students to reproduce exam conditions to maximize the effectiveness of their preparation. In short, SubjectMath.com is an excellent way for students to prepare for the GRE® test, reduce stress related to exam taking and maximize their chances of success getting admitted to their dream school.

By Prof. Amir Alexander (UCLA):

Every year thousands of aspiring mathematicians from the U.S. and beyond flood American universities with applications for graduate studies. Almost all of them are required to take the GRE® subject exam in mathematics, and the results are critical to their success. How well they do on this test could determine whether they will be admitted to the program of their choice, or even accepted at all. Yet despite the high stakes, it is nearly impossible to find study materials for the test: there are no preparation courses, and only a single sample test is provided by the exam's administrators. How is one to study for this test, which could shape one's career for decades to come? Gilad Pagi and the GP group have the answer. Their website at SubjectMath.com offers a systematic study course for the Mathematics subject exam, the only one of its kind. In dozens of clear online lectures, filled with examples and study problems, the course covers all the topics included in the exam - from the calculus to analysis to algebra and beyond. A series of practice test booklets that precisely reproduce the content, format, and conditions of the actual test accompany the course. A student that has passed through this study course will go into the exam room confident and fully-prepared, and immensely improve his or her chances of success. If you are a student planning to take the GRE® subject exam in mathematics, take notice: This is one course you cannot afford to miss.

Preface

This book is a full practice test simulating the GRE® Subject Exam in Mathematics (Graduate record examinations subject exam by Educational Testing Service - ETS). This book consists of one practice test belonging to a series of tests that are part of our online preparation course (see www.SubjectMath.com). Each test is published individually. It is highly recommended that these tests are completed in correspondence with the course as solutions may reference results and examples further discussed within the course lectures.

If you are applying for a graduate math program in the US, you must excel in this test in order to be accepted. Alas, relevant materials are scarce and you have no one to tutor you throughout the exam. That's where we, the team in GP Group, come in.

This practice test was written as a part of our preparation course, encompassing all aspects of the subject exam. The team at GP Group composed series of tests similar to the actual subject math exam in many aspects:

- ✓ The test consist of 66 multiple choice questions.
- ✓ The content of the question is taken from the official syllabus of the test.
- ✓ The style of the questions is similar to the questions in the official test example, published by ETS
- ✓ The distribution of the topics among the questions corresponds to the distribution as published by ETS and as seen in the published example test.
- ✓ The printing layout, including the space for scratch work, matches the real exam (as published).
- ✓ The test was designed to be taken in the same time frame and conditions as the real exam.
- ✓ However, all the questions are original and are not published anywhere other than with the course official materials.

Up to the date of publication, this series of books are the only practice exams not published by ETS, possessing all the above features. Considering the scarcity of the prep materials for the subject exam, this book will improve your potential score significantly and, together with the online course, provide a well rounded preparation for the test.

How to Use This Book?

Our goal is to simulate the entire experience of the actual test. Every factor of the following will make your test experience a little less stressful and more "familiar". This alone can raise your potential score. Try to to make an effort to follow the recommendations.

1. **Prepare a simulating environment:**
   - Free up at least 3 hours for taking the practice exam.
   - Prepare an empty desk - preferably, a college chair with an armrest.
   - Do not wear a watch. Make sure an analogue clock is avaiable.
   - Try to start the practice test at the same time your actual test is scheduled.
   - Prepare at least 5 sharpened number 2 pencils and an eraser.
   - Put away cellphones, water and food. Do not plan on bathroom breaks.

2. **Taking the practice exam:**
   - USE THE BOOK AS IF IT WERE YOUR EXAM BOOKLET. Use the scratch pages, make notes and draw sketches on these actual pages. This book is specifically designed for mimicking the actually test experience.
   - Print out the designated answer sheet for this exam. It can be obtained from the ETS website [1]. Mark your answers there and, later, grade your test only by looking at the answer sheet. Marking your answers correctly on the answer sheet is crucial and worth practicing.
   - Give yourself exactly 2 hours and 50 minutes, as it will be on the actual exam.
   - Grade yourself accurately - one point on any right answer. Minus 0.25 point for every wrong answer. Refer again to the above PDF file from ETS to estimate your 3 digit score (currently on page 67).

3. **Additional notes:**
   - It is important to use no.2 pencils of good quality. I recommend "Dixon Ticonderoga". Note that the "pre-sharpened" are usually not sharpened enough. These can be found easily on amazon, or any office supplies store.
   - Use a thin eraser. Otherwise, you might erase many answers on your answer sheet when attempting to correct one. I recommend "Paper Mate Tuff Stuff Eraser Stick (SN64801)".

[1] currently on: https://www.ets.org/s/gre/pdf/practice_book_math.pdf, page 69

- (Disclaimer: Those recommendations are based on my own personal experience. I do not have any relationship with these companies)

Useful Links

1. Our prep course site includes links to the different course modules, lectures and handouts. It also includes updates on new published exams, valuable lectures, promo codes for discount on our course materials. This can all be found on our course site.

   [Enter] www.SubjectMath.com

   [Enter] www.facebook.com/gpsubjectmath

2. The video lectures are published on Udemy. Look for our different "Subject Math" modules. Make sure to always check out www.SubjectMath.com for special discounts, before purchasing modules via Udemy.

   [Enter] www.udemy.com

3. Official information about the test from ETS.

   [Enter] www.ets.org/gre/subject/about/content/mathematics

4. Comments, corrections and ideas will be appreciated. Contact us at

   ↝ subjectmath@gpgroupcompany.com

About the Author

GP Group is led by Gilad Pagi. Pagi graduated 1st in class during his B.S in Math and B.S+M.S in Engineering. Pagi has more than 10 years of experience in teaching, including teaching positions in calculus and linear algebra university courses, private and group tutoring. Pagi achieved a top score in the subject math exam (900). He served as a calculus instructor at the University of Michigan, Ann Arbor, where he received his Ph.D. in Mathematics. Dr. Pagi is currently working at Google.

Acknowledgements

Special thanks to Caleb Springer, our exam "Debugger", and to Alon Ben-Haim for his enlightening comments.

Disclaimer: Although GP Group is dedicated to providing a comprehensive review of the material and quality resources for students preparing to take the GRE subject exam in mathematics, due to the nature of the exam, it may not include all the background content that might appear in the actual exam. The information given here is no replacement to the official information found on ETS site.

Before starting the practice exam, make sure to follow the instructions from the preface of the book.

If you are ready, start the exam.

Good Luck!

PRACTICE TEST

Note:

- $\log(x)$ denotes the logarithm in the natural basis.
- $\mathbb{R}, \mathbb{C}, \mathbb{Q}, \mathbb{Z}, \mathbb{N}$ denote the real numbers, complex numbers, rational numbers, integers and natural (positive) integers respectively.
- Unless specified otherwise, I is the identity matrix.
- When A is a ring, $A[x]$ denotes the polynomials with coefficients in A.
- "Such that" may be abbreviated as "s.t.".
- $\arctan(x)$ and $\tan^{-1}(x)$ denote the inverse function of $\tan(x)$, and similarly for $\sin(x)$ and $\cos(x)$.

---

1). For which plane P the following is true: $(1, 1, 1)$ is orthogonal to each line in P and the volume bounded by P and $x \geq 0, y \geq 0, z \geq 0$ is $\frac{4}{3}$?

(A) $P = \{(x, y, z) \,|\, x + y + z = 1\}$

(B) $P = \{(x, y, z) \,|\, \frac{1}{2}x + \frac{1}{2}y + \frac{1}{2}z = 1\}$

(C) $P = \{(x, y, z) \,|\, 2x + 2y + 2z = 1\}$

(D) $P = \{(x, y, z) \,|\, 3x + 3y + 3z = 1\}$

(E) $P = \{(x, y, z) \,|\, x + y + z = 0\}$

---

2). For which $a, b \in \mathbb{R}$ is $\frac{1}{a} > \frac{1}{b}$ true?

(A) $0 < b < a$

(B) $b < a < 0$

(C) $b < a$

(D) $a < b$

(E) None of the above.

USE FOR SCRATCH WORK

3). $\int_0^{\frac{\pi}{2}} \int_0^{2\cos(x)} sin(x)\, dy\, dx =$

   (A) -2

   (B) 2

   (C) -1

   (D) 1

   (E) π

4). For $x \in [1, \infty)$, $\frac{d}{dx}(2^x \log 2) = ?$

   (A) $2^x \log 2$

   (B) $2^x \log 4$

   (C) $-2^x \log 4$

   (D) $2^{x+1} \log 2$

   (E) None of the above.

USE FOR SCRATCH WORK

5). Given a function on the xy-plane, $f(x,y)$ such that $f(2,2) = 20$, and $\frac{\partial f}{\partial x} = 3x^2 + 3y$, $\frac{\partial f}{\partial y} = 3x + 2y$. Then $f(0,0) = ?$

(A) 0

(B) −1

(C) −2

(D) −3

(E) −4

6). Let $f(x)$ be the anti-derivative function of $\log(x)^2$ on the positive real line. Which of the following statement are FALSE:

   I f is strictly increasing.

   II f has a local minimum

   III f has a critical point.

(A) I

(B) II

(C) III

(D) II and III

(E) I and III

USE FOR SCRATCH WORK

7). Which of the following integrals is $\frac{\pi}{8}$?

(A) $2\int_0^{\frac{\sqrt{2}}{2}} \int_x^{\sqrt{1-x^2}} dy\, dx$

(B) $\int_0^1 \int_0^{\sqrt{1-x^2}} dy\, dx$

(C) $\int_{-1}^0 \int_0^{\sqrt{1-x^2}} dy\, dx$

(D) $\int_{-1}^0 \int_0^{\sqrt{1-y^2}} dx\, dy$

(E) None of the above.

8). $\sum_{n=1}^{\infty} \log\left((1+\frac{1}{n})^n\right) =?$

(A) 0

(B) 1

(C) e

(D) e^{-1}

(E) $+\infty$

USE FOR SCRATCH WORK

9). 10 people are present at a party. The host would like to seat 5 of them at a single round table and leave 5 of them standing. How many possibilities does the host have for such an arrangement? (Note that after people are seated in the round table, any rotation of the table is considered the same arrangement).

(A) $\binom{10}{5}$

(B) $5!\binom{10}{5}$

(C) $4!\binom{10}{5}$

(D) $(5!)^2$

(E) $5!4!$

---

10). The following is an encryption algorithm. Denote (abc) as the permutation $a \mapsto b, b \mapsto c, c \mapsto a$, and fixing all the other elements. Denote A as the following matrix:

$$\begin{pmatrix} 1 & 4 & 7 \\ 2 & 5 & 8 \\ 3 & 6 & 9 \end{pmatrix}.$$

In each step, the middle row of the matrix is used to define a permutation acting on a single digit. After each step, the first column is rotated downwards and the middle column upwards, resulting a new matrix to be used for the next step of encryption:

$$\begin{pmatrix} 3 & 5 & 7 \\ 1 & 6 & 8 \\ 2 & 4 & 9 \end{pmatrix}.$$

And so on. I.e. given a series of two digit: 2,5,8 the output is 5,5,3.

What was the input, if the output is the series 5,4,4,5,7,8,5?

(A) 5,4,4,5,7,8,5

(B) 8,4,8,8,7,3,8

(C) 2,4,3,2,7,4,2

(D) 8,4,4,5,7,1,8

(E) Such input does not exists.

USE FOR SCRATCH WORK

11). For which $x \in \mathbb{R}$ does there exist an interval (a,b) such that $x \in (a,b)$ and the function $f(x) = \arcsin(\arccos(x))$ is defined on (a,b)? (Note that $\arcsin(x) = \sin^{-1}(x)$ and $\arccos(x) = \cos^{-1}(x)$.)

(A) $x = \cos(\frac{\pi}{2})$

(B) $x = \frac{\pi}{2}$

(C) $x = \cos(1)$

(D) $x = \cos(\frac{1}{2})$

(E) $x = 1$

12). The derivative of $f(x) = \frac{\tan(3x) - \tan(x)\tan(2x)\tan(3x)}{\tan(2x) + \tan(x)}$ at $x = \frac{\pi}{7}$ is:

(A) π

(B) $\frac{\pi}{2}$

(C) 2

(D) 1

(E) 0

USE FOR SCRATCH WORK

13).
$$\lim_{x\to 1} \frac{\log(1+x)}{\sin(\pi x)} =$$

(A) 1

(B) 0

(C) ∞

(D) e

(E) Does not exists.

14). In a library, 80% of the total books are hard cover, 70% of which are also in English. Also, 70% of the total books were written before the year 2000, 80% of which are also available for sale. Choosing one book at random, which of the following is most probable?

(A) Hard cover book in English.

(B) Pre-2000 book, which is available for sale as well.

(C) Post-2000 book in English.

(D) A and B, with the same probability.

(E) There is not enough information to decide.

USE FOR SCRATCH WORK

(C) $\begin{pmatrix} 2 \\ 4 \\ 6 \end{pmatrix}$

USE FOR SCRATCH WORK

16). Which of the following is true for $x \in (1, 2)$?

(A) $0 \leq \frac{\log(x)}{x-1} \leq 0.5$

(B) $0.3 \leq \frac{\log(x)}{x-1} \leq 0.8$

(C) $0.4 \leq \frac{\log(x)}{x-1} \leq 0.9$

(D) $0.5 \leq \frac{\log(x)}{x-1} \leq 1$

(E) None of the above.

17). Let $(G, *)$ be a finite group, with identity e. Define a new binary operation on G denoted by $\otimes : G \times G \to G$, such that for all $a, b, c, d \in G$, $(a * b) \otimes (c * d) = (a \otimes b) * (c \otimes d)$, and exists $e_\otimes \in G$ such that for all $a \in G$, $a \otimes e_\otimes = e_\otimes \otimes a = a$. Which of the following is true?

I $\forall a, b \in G, a \otimes b = a * b$.

II $e = e_\otimes$.

III G is abelian.

(A) I

(B) II and III

(C) I and II

(D) I and III

(E) I, II and III

USE FOR SCRATCH WORK

18). Which of the following is true about a group of order 6?

(A) It is Abelian.

(B) It is cyclic

(C) It has a member of order 2.

(D) It has a member of order 4.

(E) None of the above.

19).
$$\lim_{n \to \infty} \inf_{x \in \mathbb{R}} (e^x - nx) = ?$$

(A) $+\infty$

(B) $-\infty$

(C) e

(D) $\frac{1}{e}$

(E) Does not exists.

USE FOR SCRATCH WORK

20). Given $\forall x \in \mathbb{R}$, $f(x) = f(x^2)$, what is true about $f(x)$?

(A) $f(x)$ is bounded.

(B) If $f(x)$ is a polynomial, $f(x)$ has a bounded derivative.

(C) If $f(x)$ is a polynomial, $f(x) = 0$.

(D) $f(x)$ is odd.

(E) Such $f(x)$ does not exists.

---

21).
$$\int_0^{\frac{\pi}{2}} |\cos(x)| + \int_{\frac{\pi}{2}}^{\pi} |\sin(x)| + \int_{\pi}^{\frac{3\pi}{2}} |\cos(x)| + \int_{\frac{3\pi}{2}}^{2\pi} |\sin(x)| = ?$$

(A) 1

(B) 2

(C) 3

(D) 4

(E) π

USE FOR SCRATCH WORK

22). Let $f(x) = x^{-a}, g(x) = x^{-b}, a, b \in (0, 10)$. Which of the following is true?

(A) $f(x) > g(x)$ when $x \in (3, 6)$ and $a > b > 1$.

(B) $f(x) < g(x)$ when $x \in (0, 4)$ and $a > b$.

(C) $f(x) > g(x)$ when $x \in (0.5, 0.9)$ and $a < b$.

(D) $f(x) < g(x)$ when $x \in (0.3, 0.5)$ and $a < b < 1$.

(E) None of the above.

---

23). Suppose we have an algorithm B that is given as input two **sorted** lists of numbers l_1, l_2 and outputs a **sorted** list of numbers l_3 composed of l_1 and l_2, i.e $l_3 = B(l_1, l_2)$. Consider an algorithm A for sorting a list of numbers, when the an input is a single list of numbers l(not necessarily sorted), and the output $\tilde{l}$ is a sorted list of numbers composed of the numbers in l, i.e. $\tilde{l} = A(l)$. The following is pseudo-code describing A:

1. get input l.

2. if l is sorted, output l.

3. else,

    3.1 separate l in the middle into two smaller lists, the left part l_1 and the right part l_2, where $0 \leq length(l_1) - length(l_2) \leq 1$.

    3.2 calculate $A(l_1)$, store result in x

    3.3 $l_3 = x$

    3.4 calculate $A(l_2)$, store result in y

    3.5 $l_4 = y$

    3.6 output $B(l_3, l_4)$

Note that a list of length 1 is sorted. For an input $l = [3, 1, 2, 4]$, what would be the content of l_3 for each time the algorithm A executes step 3.3.

(A) $l_3 = [1, 3]$

(B) $l_3 = [3, 1]$

(C) $l_3 = [1, 3], l_3 = [1]$

(D) $l_3 = [1, 3], l_3 = [3]$

(E) $l_3 = [3], l_3 = [1, 3]$

USE FOR SCRATCH WORK

24). For a real function $f(x)$, Let $A_f = \{x \mid f(x) \neq 0\}$ be a subset of $\mathbb{R}$ on which $f(x)$ is non zero. For each positive integer n, let $B_n = \{\frac{m}{n} \mid m \in \mathbb{Z}, \} \cap [0,1]$, Define a map G on real functions by $G(f) = \limsup_{n \to \infty} \#(A_f \cap B_n)$, where $\#(C)$ is the number of elements in the set C if C is of a finite cardinality, and $\#(C) = \infty$ otherwise. Which of the following is FALSE?

(A) Exists $f(x)$ such that $G(f) = 0$.

(B) Exists a non constant $f(x)$ such that $G(f) = 0$.

(C) Exists a continuous $f(x)$ such that $G(f) = 0$.

(D) Exists $f(x)$ such that the series $\#(A_f \cap B_n)$ does not have a limit.

(E) Exists a prime integer p, such that for all $f(x)$, $G(f) \neq p$.

25). Let a, b be integers such that $5a + 9b$ is divisible by 17. Which of the following is not guaranteed to be divisible by 17?

(A) $2a + 2b$

(B) $8a + 11b$

(C) $11a + 13b$

(D) $13a + 3b$

(E) $a + 12b$

USE FOR SCRATCH WORK

26). Let $f : (1, \infty) \to [0, \infty)$ be a function, such that the improper integral $\int_1^\infty f(x)\, dx$ converges. Which of the followings true.

   I $\lim_{x \to \infty} f(x)$ exists.

   II If $f(x)$ is monotonous decreasing, then $\lim_{x \to \infty} f(x)$ exists.

   III If $f(x)$ has derivatives of every order then $\lim_{x \to \infty} f(x)$ exists.

   (A) I

   (B) II

   (C) III

   (D) II and III

   (E) I, II and III

27). Let $A = \{(x, \sin(\frac{1}{x})) \,|\, x \in (0, 1)\} \cup \{0\} \times [0, 1]$ be a subset of $\mathbb{R}^2$. Which of the following is true about A?

   I A is path connected.

   II A is connected.

   III A is closed.

   (A) I, II

   (B) II

   (C) III

   (D) II and III

   (E) I, II and III

USE FOR SCRATCH WORK

28). Let:
$$f(x) = \begin{cases} \arctan(x) & |x| < 1 \\ ax^3 + bx^2 + cx + d & |x| \geq 1 \end{cases},$$

where $\arctan(x) = \tan^{-1}(x)$. For which a, b, c, d does $f(x)$ have continuous second derivative?

(A) $a = -\frac{1}{3}, b = 0, c = 1, d = 0$

(B) $a = 1, b = 0, c = 1, d = -2 + \frac{\pi}{4}$

(C) $a = 0, b = \frac{\pi}{4}, c = 0, d = 0$

(D) $a = -\frac{1}{12}, b = 0, c = \frac{3}{4}, d = 0$

(E) None of the above.

---

29). Consider the hyperbola $\frac{x^2}{a^2} - \frac{y^2}{b^2} = 1$. Let $(\pm c, 0)$ be the foci ($c = \sqrt{a^2 + b^2}$). Let C be the circle centered at the origin going through the foci. Denote A as the area of the circular sector of C bounded by the x-axis and the hyperbola's asymptote in the first quadrant. For which hyperbola, $A = \pi$?

(A) $\frac{x^2}{9} - \frac{y^2}{9} = 1$

(B) $\frac{x^2}{3} - \frac{y^2}{3} = 1$

(C) $\frac{x^2}{9} - \frac{y^2}{3} = 1$

(D) $\frac{x^2}{3} - \frac{y^2}{9} = 1$

(E) None of the above.

USE FOR SCRATCH WORK

30). Let:
$$A = \int_1^e \frac{\log(x)}{x}\,dx.$$

Which of the following is true?

(A) $0 \leq A < 0.2$

(B) $0.2 \leq A < 0.5$

(C) $0.5 \leq A < 0.7$

(D) $0.7 \leq A < 1$

(E) None of the above.

---

31).
$$\lim_{x \to 1} \frac{(1-x)^3 + 3x - 3}{3\arctan(x-1)} = ?$$

(A) 0

(B) 1

(C) $\frac{1}{3}$

(D) ∞

(E) None of the above.

USE FOR SCRATCH WORK

(A) $T: M_3(\mathbb{R}) \to \mathbb{R}_5[x]$

USE FOR SCRATCH WORK

33). Observe the following graphs of $\arctan(x), \log(x)$ and $\frac{1}{x}$:

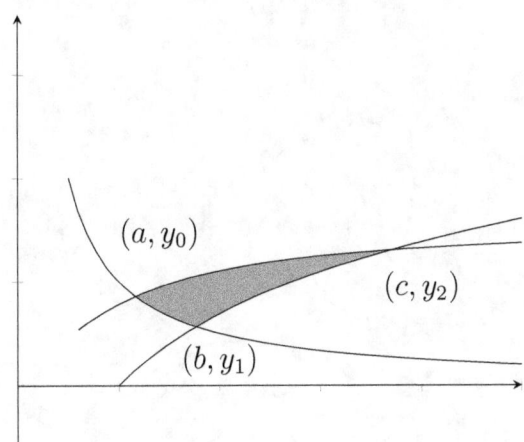

$(a, y_0), (b, y_1), (c, y_2)$ are the coordinates of the intersection points as seen in the figure above. The shaded region is calculated by the following integral($a \leq b \leq c$):

$$\int_a^b \int_{f(x)}^{g(x)} dy\, dx + \int_b^c \int_{h(x)}^{q(x)} dy\, dx$$

Which of the following is true?

I $a \cdot \arctan(a) = b \cdot \log(b)$

II $h(x)$ may be equal to $\log(x)$

III $g(x)$ may be equal to $q(x)$

(A) I

(B) II

(C) III

(D) I and II

(E) I, II and III

USE FOR SCRATCH WORK

34). Let $\mathbb{C}^4$ denote the 4-dimensional vector space of 4×1 column vectors over $\mathbb{C}$. Let V be the intersection of all the subspaces of $\mathbb{C}^4$. Which of the following is true?

   I   If V is a vector space, then $\dim V = 0$.

   II The cardinality of V is 1.

   III If V is a vector space, then the empty set $\emptyset$ is a basis of V.

   (A) I

   (B) II

   (C) III

   (D) I and II

   (E) I, II and III

35). Observe the following matrix:

$$\begin{pmatrix} i & i & 0 & i & 0 & i \\ 0 & 1 & -1 & -1 & 2 & 2 \\ 1 & 0 & 1 & 2 & -2 & -1 \end{pmatrix}$$

What is its row vectors space dimension?

   (A) 2

   (B) 3

   (C) 4

   (D) 5

   (E) 6

USE FOR SCRATCH WORK

36). Let E be the part of the ellipse $x^2 + \frac{y^2}{4} = 1$ in the first quadrant. Let A be a rectangle with a vertex at the origin and an opposing vertex on E. What is the maximum possible value of the area of such A?

(A) $\frac{1}{2}$

(B) 1

(C) $\sqrt{2}$

(D) $\frac{\sqrt{2}}{2}$

(E) None of the above

37). Let E be the graph of $y = \frac{1}{\sqrt{x}}$. Let A be a rectangle with a vertex at the origin and an opposing vertex on E. What is the maximum possible area of such a rectangle A?

(A) $\frac{1}{2}$

(B) 1

(C) $\sqrt{2}$

(D) $\frac{\sqrt{2}}{2}$

(E) None of the above

USE FOR SCRATCH WORK

38). Consider the following argument:

   a) Let the real sequences $a_n \to L$ and $b_n \to L$, as $n \to \infty$. Then,

   b) $a_n - b_n \to 0$, as $n \to \infty$. Then,

   c) For all $\epsilon > 0$, exists a sub-sequence n_k such that $\sum_{k=1}^{\infty} |a_{n_k} - b_{n_k}| < \epsilon$. Then

   d) Exists a sub-sequence n_k such that $a_{n_k} = b_{n_k}$

Which of the following is true?

   I a) does not imply b).

   II b) does not imply c).

   III c) does not imply d).

(A) I

(B) II

(C) III

(D) II and III

(E) I and III

USE FOR SCRATCH WORK

39). Let A be the following matrix:
$$\begin{pmatrix} 1 & 2 \\ 2 & 1 \end{pmatrix}$$

Which of the following is A^5?

(A) $\begin{pmatrix} 1 & 2 \\ 2 & 1 \end{pmatrix}$

(B) $\begin{pmatrix} 121 & 122 \\ 122 & 121 \end{pmatrix}$

(C) $\begin{pmatrix} 121 & 2 \\ 2 & 121 \end{pmatrix}$

(D) $\begin{pmatrix} 212 & 121 \\ 121 & 212 \end{pmatrix}$

(E) $\begin{pmatrix} 112 & 221 \\ 221 & 130 \end{pmatrix}$

USE FOR SCRATCH WORK

40). Let $f(x) = \int_1^{x^2} \frac{\sin(x)}{t} \, dt$.
$$\lim_{x \to 1} \frac{f(x)}{x-1} = ?$$

(A) 0

(B) 1

(C) $\sin(1)$

(D) $2\sin(1)$.

(E) None of the above.

---

41). Let $f(x, y) = 0$ be a solution of the differential equation $3y(1+x)\,dx + 3x\,dy = 0$ such that $y = 1$ when $x = 1$. What is y when $x = 2$?

(A) 0

(B) 1

(C) e

(D) $6e^2$

(E) None of the above.

USE FOR SCRATCH WORK

42). The following represents a portion of the graph of the **derivative** of $f(x)$, on the interval (a, b):

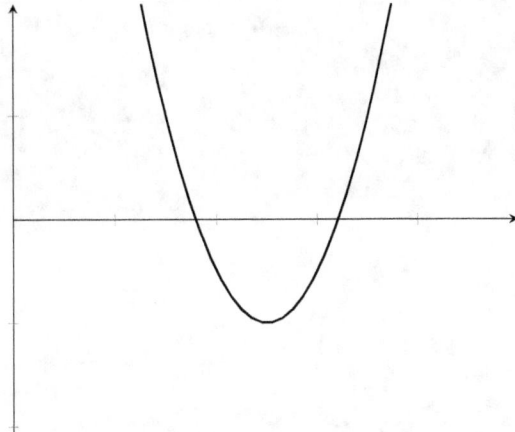

Which of the following may represent the graph of $f(x)$ on the interval (a, b)?

(A) 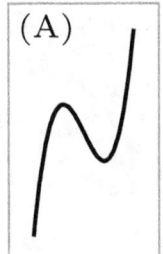 (B) 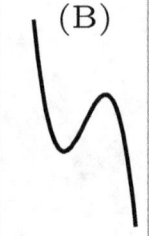 (C) 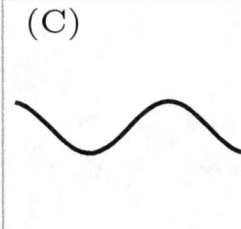 (D)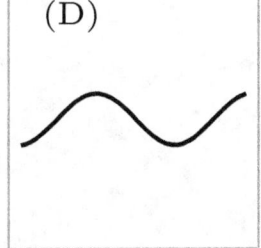

(E)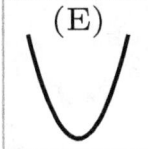

USE FOR SCRATCH WORK

43). Which of the following real matrices, has i and $1+i$ as eigenvalues?

(A) $A = \begin{pmatrix} 8 & 0 & 4 & 8 \\ 4 & -12 & -8 & 0 \\ -2 & 20 & 14 & 4 \\ -3 & -8 & -7 & -2 \end{pmatrix} \cdot \begin{pmatrix} \frac{1}{4} & 0 & 0 & 0 \\ 0 & \frac{1}{4} & 0 & 0 \\ 0 & 0 & \frac{1}{4} & 0 \\ 0 & 0 & 0 & \frac{1}{4} \end{pmatrix}$

(B) $B = \begin{pmatrix} 1 & 4 & -1 & 0 \\ 2 & 8 & -2 & 0 \\ 5 & -4 & 3 & 0 \\ 5 & 0 & -5 & -4 \end{pmatrix} \cdot \begin{pmatrix} \frac{1}{4} & 0 & 0 & 0 \\ 0 & \frac{1}{4} & 0 & 0 \\ 0 & 0 & \frac{1}{4} & 0 \\ 0 & 0 & 0 & \frac{1}{4} \end{pmatrix}$

(C) $C = \begin{pmatrix} 1 & 4 & -1 & 0 \\ -5 & 9 & -1 & 0 \\ 5 & -4 & 3 & 0 \\ 5 & 0 & -5 & -4 \end{pmatrix} \cdot \begin{pmatrix} \frac{1}{4} & 0 & 0 & 0 \\ 0 & \frac{1}{4} & 0 & 0 \\ 0 & 0 & \frac{1}{4} & 0 \\ 0 & 0 & 0 & \frac{1}{4} \end{pmatrix}$

(D) $D = \begin{pmatrix} 1 & 4 & -1 & 0 \\ -5 & 8 & -1 & 0 \\ 5 & -4 & 3 & 0 \\ 5 & 0 & -5 & -5 \end{pmatrix} \cdot \begin{pmatrix} \frac{1}{4} & 0 & 0 & 0 \\ 0 & \frac{1}{4} & 0 & 0 \\ 0 & 0 & \frac{1}{4} & 0 \\ 0 & 0 & 0 & \frac{1}{4} \end{pmatrix}$

(E) $E = \begin{pmatrix} 1 & 4 & -1 & 0 \\ -5 & 8 & -1 & 0 \\ 5 & -4 & 4 & 0 \\ 5 & 0 & -5 & -4 \end{pmatrix} \cdot \begin{pmatrix} \frac{1}{4} & 0 & 0 & 0 \\ 0 & \frac{1}{4} & 0 & 0 \\ 0 & 0 & \frac{1}{4} & 0 \\ 0 & 0 & 0 & \frac{1}{4} \end{pmatrix}$

USE FOR SCRATCH WORK

44). Let $M_3(\mathbb{R})$ be the vector space of 3×3 matrices over $\mathbb{R}$. Which of the following is true?

  I Exists a basis of $M_3(\mathbb{R})$ which contains no matrices with zero trace.

  II Exists a basis of $M_3(\mathbb{R})$ which contains exactly 8 matrix with zero trace.

  III Exists a basis of $M_3(\mathbb{R})$ which contains exactly 9 matrix with zero trace

(A) I

(B) II

(C) I,II

(D) II,III

(E) I, II and III

---

45). Let $f(z) = \frac{1}{z^2 - a^2}$ be a complex function with a real parameter $a \in \mathbb{R}$. Let C be the circle $|z| < R$, $R \in \mathbb{R}$, $R > 0$ oriented counterclockwise. For which choice of R and a is the integral $\oint_C f(z)\, dz$ non-zero (as long as C does not intersect $(\pm a, o)$)?

(A) Any R and a, as long as C does not intersect $(\pm a, o)$.

(B) Any R, $a = 0$.

(C) $a > 0, R > a$.

(D) $a > 0, R < a$.

(E) Such choice is not possible.

USE FOR SCRATCH WORK

46). **(A) 1**

Partial fractions: $\frac{1}{z(z-1)(z-2)} = \frac{1/2}{z} - \frac{1}{z-1} + \frac{1/2}{z-2}$. For $|z|>2$, expanding in powers of $1/z$ gives coefficient of z^{-n} equal to $-1 + 2^{n-2}$ for $n\geq 2$ (and $1/2-1+1/2=0$ for $n=1$). This is zero for $n=1,2$ and equals 1 for $n=3$.

47). **(D)** When the algorithm stops, the gambler has always netted a profit of exactly \$1. The probability that the algorithm never stops is 0 (since $(19/37)^\infty = 0$), so with probability $1 \geq 0.8$ the gambler profits at least \$1.

USE FOR SCRATCH WORK

48). Let $f(z) = \sum_{n=1}^{\infty} (-1)^{n-1} \frac{1}{4^n} z^{2n-1}$. Denote $A = \{z \mid z \text{ is a pole of } f(z)\}$ and $r = \inf_A\{|z|\}$. Let z_0 be a pole of $f(z)$ such that $|z_0| = r$, if such z_0 exists. What is a possible value of z_0?

(A) -4

(B) i

(C) $1+i$

(D) $-2i$

(E) Such z_o does not exists.

---

49). For $b, c \in \mathbb{Z}$, let $A_{b,c} = \{bx + cy \mid x, y \in \mathbb{Z}\}$ be a subset of the ring $\mathbb{Z}$. We say that V is a "generated" subset of $\mathbb{Z}$ if $\exists v \in V$ such that $\forall w \in V, \exists n \in \mathbb{Z}$ for which $w = nv$. For which of the following pairs (b, c) the set $A_{b,c}$ is <u>NOT</u> generated?

(A) $(1, 2)$

(B) $(2, 3)$

(C) $(16, 48)$

(D) $(540, 448)$

(E) None.

USE FOR SCRATCH WORK

50). 35 people are attending a party. Any pair of people may or may not talk during the night. At the end of the night, each participant will have conversed with 0,1,2,...,33 or 34 different people. Let p be the probability that there exist 2 different people which have conversed with the same number of people. What is true about p?

(A) $0 \leq p < 0.2$

(B) $0.2 \leq p < 0.4$

(C) $0.4 \leq p < 0.6$

(D) $0.6 \leq p < 0.8$

(E) $0.8 \leq p \leq 1$

51). Which of the following is of finite or countable cardinality?

I $\mathbb{Q} \times \mathbb{Q} \times \mathbb{Q}$.

II $\bigcup_{n \in \mathbb{N}} (\mathbb{Z} + \frac{1}{n})$ where $\mathbb{Z} + a = \{x + a \,|\, x \in \mathbb{Z}\}$.

III All the subsets of $\mathbb{N}$.

(A) I

(B) I,II

(C) II,III

(D) I,III

(E) I,II and III

USE FOR SCRATCH WORK

52). Denote $\mathbb{C}_4[x]$ to be the vector space spanned by the polynomials of degree ≤ 4 with coefficients in $\mathbb{C}$. Let $D : \mathbb{C}_4[x] \to \mathbb{C}_4[x]$ be the derivative linear transformation, i.e. $D(f(x)) = \frac{df(x)}{dx}$. What is true about D? (denote tr as the trace, det as the determinant).

I $\operatorname{tr}(D) > 3$

II $\operatorname{tr}(D) = \det(D)$

III The characteristic polynomial of D is $x^5 + x^3 + x$.

(A) I

(B) II

(C) III

(D) II,III

(E) I,II and III

---

53). For every Lebesgue measurable set A denote $\mu(A)$ as the Lebesgue measure of A. Let $f(x)$ be the following function:

$$f(x) = \begin{cases} x & x \in \mathbb{Q} \\ 0 & \text{otherwise} \end{cases}.$$

Which of the following is true?

I $f(x)$ is a Lebesgue measurable function.

II $\int_\mathbb{R} f(x)\, d\mu < \infty$

III $\int_\mathbb{R} f(x)\, d\mu$ is equal to the (improper) Riemann integral of $f(x)$.

(A) II

(B) I,II

(C) I,III

(D) II,III

(E) I,II and III

USE FOR SCRATCH WORK

54). Let G be the Dihedral group of order 8. Let n to be the number of elements in G of order less than 3. How many groups of order n exists (up to isomorphism)?

(A) 0

(B) 1

(C) 2

(D) 3

(E) 6

---

55). Let $f(x,y) = x^3 + ay^2 + 4xy$, where $a \neq 0$. Which of the following must be true?

(A) $\exists a \in [2,6]$ such that $f(x,y)$ has no saddle points.

(B) $\exists a \in [-2,0)$ such that $f(x,y)$ has 2 distinct minimum points.

(C) $\exists a \in [1,5]$ such that $f(x,y)$ has one maximum point.

(D) $\exists a \in [-2,6]$ such that $f(x,y)$ has 2 distinct maximum points.

(E) $\exists a \in [3,7]$ such that $f(x,y)$ has exactly one saddle point.

USE FOR SCRATCH WORK

56). Let $f_n(x) : [0,1] \to \mathbb{R}$ be a sequence of functions. Which of the following must be true? (Note: a statement in true almost everywhere on an interval I means that the set of all $x \in I$ for which the statement is not true, has a Lebesgue measure zero. Also note that the integrals may be improper).

I $\lim_{n \to \infty} \int_0^1 |f_n(x)| dx = 0$ implies $f_n(x) \to 0$ almost everywhere on $[0,1]$

II $\lim_{n \to \infty} \int_0^1 f_n(x) dx < \infty$ DOES NOT imply that $\exists N$ such that $f_n(x)$ is bounded on $[0,1]$ for all $n > N$.

III If $f_n(x)$ converges pointwise to a continuous function $f(x)$ then $\exists N$ such that $f_n(x)$ is bounded on $[0,1]$ for all $n > N$.

(A) I

(B) II

(C) I,II

(D) II,III

(E) I,II and III

---

57). Let L be the approximation of $\log(x)$ for $x = 1.02$ by its Taylor expansion of order 2 around 1. Denote $r = \log(1.02) - L$. Which of the following is true?

(A) $0.0001 \leq r \leq 0.001$

(B) $-0.01 \leq r \leq -0.001$

(C) $0.00001 \leq r \leq 0.0001$

(D) $-0.001 \leq r \leq -0.0001$

(E) $0.000001 \leq r \leq 0.00001$

USE FOR SCRATCH WORK

58). Let $f(z): X \to \mathbb{C}$ be an analytic function, $X \subset \mathbb{C}$ such that $|f(z)|$ is constant on a non-empty open set of the complex plane, contained in X. Which of the following is true about $f(z)$? (Note: $f(z) = f(x+iy)$, $x,y \in \mathbb{R}$, $f_x(z)$ is the partial derivative of $f(z)$ with respect to x, and similar for y).

(A) $f_x(z)$ and $f_y(z)$ are equal for some subset of the complex plane of infinite cardinality.

(B) $f(z)$ must be unbounded.

(C) $f(z)$ must be constant.

(D) Exists a point $z_0 \in \mathbb{C}$ such that the power series expansion of $f(z)$ around z_0 converges to $f(z)$ for any z such that $|z - z_0| < 1$.

(E) Such $f(z)$ does not exists.

---

59). A fair coin is tossed 10000 times. Let $p_{a,b}$ be the probability that the number of heads H satisfy $a \leq H \leq b$. Which of the following a,b pairs would ensure that $0.37 \leq p_{a,b} \leq 0.45$

(A) $a = 5000, b = 7000$

(B) $a = 5050, b = 5100$

(C) $a = 4975, b = 5025$

(D) $a = 4950, b = 5050$

(E) None of the above

USE FOR SCRATCH WORK

60). The following data points are given: (1,1), (2,3), (4,5), (2,3), (4,5), (5,1). What is the linear regression line equation (minimizing the square distance between the line and the data points)?

(A) $y = \frac{1}{2}x + 1$

(B) $y = \frac{1}{3}x + 1$

(C) $y = \frac{1}{2}x + 2$

(D) $y = \frac{1}{3}x + 2$

(E) $y = \frac{1}{2}x + 2.5$

---

61). Denote:
$$A = \oint_C ye^{xy}dx + (xe^{xy} + x)dy.$$

What is the value of A, where C is the graph in the following figure, oriented counterclockwise?

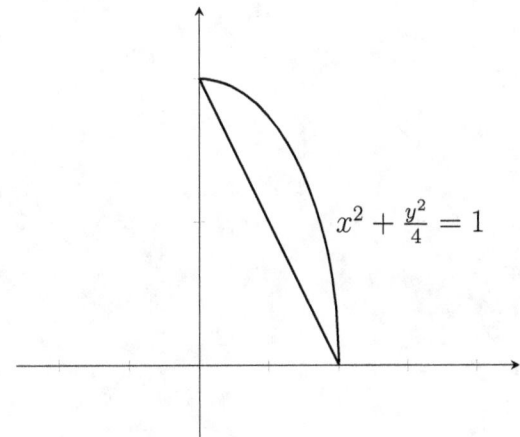

(A) $\pi - 1$

(B) $\pi - 2$

(C) $2\pi - 1$

(D) $2\pi - 2$

(E) None of the above.

USE FOR SCRATCH WORK

62). **D) 7**

63). **C) I, III**

USE FOR SCRATCH WORK

64). How many abelian groups of order 1200 are not cyclic (up to isomorphism)?

(A) 1

(B) 2

(C) 5

(D) 9

(E) 10

---

65). Which of the following is FALSE?

(A) $\mathbb{R}$ is a vector space over $\mathbb{Q}$.

(B) $\{a + b\sqrt{7} \,|\, a, b \in \mathbb{Q}\}$ is a field.

(C) The set $\{0, 1, 2, 3, 4\}$ with addition modulo 5 and multiplication modulo 5 is a ring.

(D) All the invertible elements the real 2 by 2 matrices is a group.

(E) None of the above is false.

USE FOR SCRATCH WORK

66). Which of the following can be a continuous function?

(A) Bounded and surjective (onto) function $f : [0,1] \to \mathbb{R}$

(B) unbounded function $f : [0,1] \to \mathbb{R}$

(C) Surjective(onto) function $f : [0,1] \to (0,1)$

(D) One-to-one function $f : [0,1] \to (0,1)$

(E) None of the above.

END OF PRACTICE TEST

Answers and Solutions

1). For which plane P the following is true: $(1,1,1)$ is orthogonal to each line in P and the volume bounded by P and $x \geq 0, y \geq 0, z \geq 0$ is $\frac{4}{3}$?

   (A) $P = \{(x,y,z) \,|\, x+y+z = 1\}$

   (B) $P = \{(x,y,z) \,|\, \frac{1}{2}x + \frac{1}{2}y + \frac{1}{2}z = 1\}$

   (C) $P = \{(x,y,z) \,|\, 2x + 2y + 2z = 1\}$

   (D) $P = \{(x,y,z) \,|\, 3x + 3y + 3z = 1\}$

   (E) $P = \{(x,y,z) \,|\, x+y+z = 0\}$

   The answer is B. The volume is in fact the volume of a pyramid with a triangle base of area $2 \times 2/2 = 2$ and height of 2, so the volume, by the pyramid formula, is $V = h \times base/3 = 4/3$. We could integrate to solve this question, but it takes more time. Notice that all the planes has a normal in the direction of $(1, 1, 1)$.

2). For which $a, b \in \mathbb{R}$ is $\frac{1}{a} > \frac{1}{b}$ true?

   (A) $0 < b < a$

   (B) $b < a < 0$

   (C) $b < a$

   (D) $a < b$

   (E) None of the above.

   Answer is E. Drawing the graph of $\frac{1}{x}$ make it easy to see that none of the answers guarantee $\frac{1}{a} > \frac{1}{b}$. Note that the set of (a,b)-pairs in A and B is a subset of the set of (a,b)-pairs in C. Hence we can eliminate C (If C is true, A must be true).

3). $\int_0^{\frac{\pi}{2}} \int_0^{2\cos(x)} \sin(x) \, dy \, dx =$

   (A) -2

   (B) 2

   (C) -1

   (D) 1

   (E) π

   Answer is D. Note that we can eliminate all negative answers after noticing the equivalence to $\int_0^{\frac{\pi}{2}} \sin(2x)\,dx$

$$\int_0^{\frac{\pi}{2}} \int_0^{2\cos(x)} \sin(x)\,dy\,dx = \int_0^{\frac{\pi}{2}} 2\cos(x)\sin(x)\,dx$$

$$= \int_0^{\frac{\pi}{2}} \sin(2x)\,dx = -\frac{1}{2}\cos(2x)\Big|_0^{\frac{\pi}{2}} = -\frac{1}{2}(-1-1) = 1.$$

4). For $x \in [1, \infty)$, $\frac{d}{dx}(2^x \log 2) =$?

(A) $2^x \log 2$

(B) $2^x \log 4$

(C) $-2^x \log 4$

(D) $2^{x+1} \log 2$

(E) None of the above.

Answer is E. The right answer is $\frac{d}{dx}(2^x \log 2) = (\log 2)^2 2^x$. Notice: $\log 2^2 = \log 4 = 2 \log 2 \neq (\log 2)^2$.

5). Given a function on the xy-plane, $f(x, y)$ such that $f(2, 2) = 20$, and $\frac{\partial f}{\partial x} = 3x^2 + 3y$, $\frac{\partial f}{\partial y} = 3x + 2y$. Then $f(0, 0) =$?

(A) 0

(B) -1

(C) -2

(D) -3

(E) -4

Answer is E. Integrate with respect to x: $f(x, y) = x^3 + 3xy + g(y)$, then take derivative with respect to y: $f_y = 3x + g'(y) = 3x + 2y \Rightarrow g(y) = y^2 + C \Rightarrow f = x^3 + 3xy + y^2 + C$. Plugging in the given point gives us $C = -4$ and the rest follows.

6). Let $f(x)$ be the anti-derivative function of $\log(x)^2$ on the positive real line. Which of the following statement are FALSE:

I f is strictly increasing.

II f has a local minimum

III f has a critical point.

(A) I

(B) II

(C) III

(D) II and III

(E) I and III

The answer is B. The derivative is always positive except for a single point $x = 1$ which is the (single) critical point. Thus, the graph is always increasing even in a small neighborhood around $x = 1$. The first step to solve it is to draw the following picture of $\log(x)$, and $|\log(x)|$ which behaves similar to $\log(x)^2$.

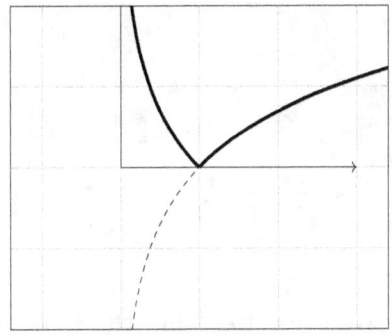

7). Which of the following integrals is $\frac{\pi}{8}$?

(A) $2 \int_0^{\frac{\sqrt{2}}{2}} \int_x^{\sqrt{1-x^2}} dy\, dx$

(B) $\int_0^1 \int_0^{\sqrt{1-x^2}} dy\, dx$

(C) $\int_{-1}^0 \int_0^{\sqrt{1-x^2}} dy\, dx$

(D) $\int_{-1}^0 \int_0^{\sqrt{1-y^2}} dx\, dy$

(E) None of the above.

Answer E. Each of the integrals represent a different area calculation of a quarter of the unit circle. Calculation gives $\frac{\pi}{4}$ for each answer. Notice that A is a calculation of twice the area of a one eighth of the circle. By drawing the boundaries, one can get the correct answer. Calculating the integral directly is a waist of precious time.

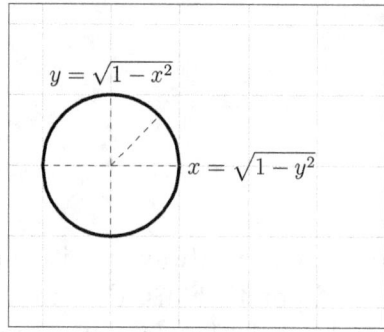

8). $\sum_{n=1}^{\infty} \log\left((1+\frac{1}{n})^n\right) =?$

(A) 0

(B) 1

(C) e

(D) e^{-1}

(E) $+\infty$

Answer is E. This is a positive series, and the limit of the summands is not 0. Thus the series diverges to $+\infty$

9). 10 people are present at a party. The host would like to seat 5 of them at a single round table and leave 5 of them standing. How many possibilities does the host have for such an arrangement? (Note that after people are seated in the round table, any rotation of the table is considered the same arrangement).

(A) $\binom{10}{5}$

(B) $5!\binom{10}{5}$

(C) $4!\binom{10}{5}$

(D) $(5!)^2$

(E) $5!4!$

Answer is C. After choosing 5 out of 10, we need to arrange 5 people in a round table. Fixing one person eliminates the rotation and leaves us with 4! possibilities.

10). The following is an encryption algorithm. Denote (abc) as the permutation $a \mapsto b, b \mapsto c, c \mapsto a$, and fixing all the other elements. Denote A as the following matrix:

$$\begin{pmatrix} 1 & 4 & 7 \\ 2 & 5 & 8 \\ 3 & 6 & 9 \end{pmatrix}.$$

In each step, the middle row of the matrix is used to define a permutation acting on a single digit. After each step, the first column is rotated downwards and the middle column upwards, resulting a new matrix to be used for the next step of encryption:
$$\begin{pmatrix} 3 & 5 & 7 \\ 1 & 6 & 8 \\ 2 & 4 & 9 \end{pmatrix}.$$

And so on. I.e. given a series of two digit: 2,5,8 the output is 5,5,3.

What was the input, if the output is the series 5,4,4,5,7,8,5?

(A) 5,4,4,5,7,8,5

(B) 8,4,8,8,7,3,8

(C) 2,4,3,2,7,4,2

(D) 8,4,4,5,7,1,8

(E) Such input does not exists.

Answer is C. It can be checked directly, but the fastest way to solve this is the following. Observe that the algorithm repeats itself every 3 digits, meaning, that the digits in the 1,4,7-places must all be the same and the "preimage" of the digit 5 must populate those places. This preimage is 2, by direct observation, or by using the given example. The input must exists since the algorithm is reversible.

11). For which $x \in \mathbb{R}$ does there exist an interval (a, b) such that $x \in (a, b)$ and the function $f(x) = \arcsin(\arccos(x))$ is defined on (a, b)? (Note that $\arcsin(x) = \sin^{-1}(x)$ and $\arccos(x) = \cos^{-1}(x)$.)

(A) $x = \cos(\frac{\pi}{2})$

(B) $x = \frac{\pi}{2}$

(C) $x = \cos(1)$

(D) $x = \cos(\frac{1}{2})$

(E) $x = 1$

Answer is D. $\arcsin(y)$ defined on $y \in [-1, 1]$, so $-1 \leq \arccos(x) \leq 1$. Drawing the graph, we see that $\cos(1) \leq x \leq 1$, so in order to be defined on an interval around x, $x \in (\cos(1), 1)$. Only D is such x. Notice that by the Taylor's expansion, $\cos(1) \approx 1 - \frac{1^2}{2} = \frac{1}{2}$.

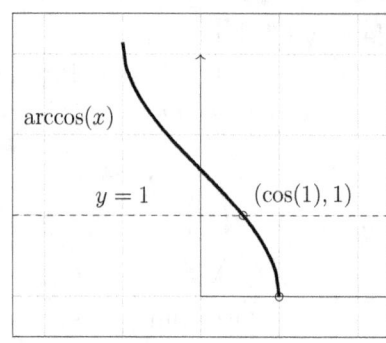

12). The derivative of $f(x) = \frac{\tan(3x) - \tan(x)\tan(2x)\tan(3x)}{\tan(2x) + \tan(x)}$ at $x = \frac{\pi}{7}$ is:

(A) π

(B) $\frac{\pi}{2}$

(C) 2

(D) 1

(E) 0

Answer is E. Using the trigonometric equivalence for $\tan(a+b)$ we can see that the function is constant 1.

$$f(x) = \tan(3x)\left(\frac{1 - \tan(2x)\tan(x)}{\tan(2x) + \tan(x)}\right) = \frac{\tan(3x)}{\tan(3x)} = 1$$

13).
$$\lim_{x \to 1} \frac{\log(1+x)}{\sin(\pi x)} =$$

(A) 1

(B) 0

(C) ∞

(D) e

(E) Does not exists.

Answer is E. The limit of the numerator is $\log(2)$ and the limit of the denominator is 0, but $\sin(x)$ changes signs around π so in fact the limit is $+\infty$ from the left, and $-\infty$ from the right.

14). In a library, 80% of the total books are hard cover, 70% of which are also in English. Also, 70% of the total books were written before the year 2000, 80% of which are also available for sale. Choosing one book at random, which of the following is most probable?

(A) Hard cover book in English.

(B) Pre-2000 book, which is available for sale as well.

(C) Post-2000 book in English.

(D) A and B, with the same probability.

(E) There is not enough information to decide.

Answer is D. Both descriptions in A and B, describe a total of 56% of the books in the library. In C, we describe less than 30% of the books, since the books were written after the year 2000.

15). Consider a linear transformation $T : \mathbb{R}^3 \to \mathbb{R}^3$ which has 4 as an eigenvalue with eigenvector $\begin{pmatrix} 1 \\ 2 \\ 3 \end{pmatrix}$. Given $T\begin{pmatrix} 1 \\ 0 \\ 1 \end{pmatrix} = \begin{pmatrix} 1 \\ 2 \\ 3 \end{pmatrix}$, what is $T\begin{pmatrix} -1 \\ 2 \\ 1 \end{pmatrix} = ?$

(A) $\begin{pmatrix} 4 \\ 0 \\ 4 \end{pmatrix}$

(B) $\begin{pmatrix} 2 \\ 4 \\ 3 \end{pmatrix}$

(C) $\begin{pmatrix} 2 \\ 4 \\ 6 \end{pmatrix}$

(D) $\begin{pmatrix} 4 \\ 8 \\ 12 \end{pmatrix}$

(E) There is not enough information, since $\mathbb{R}^3$ is of dimension 3.

Answer is C. Denote v as the eigenvector, and $w = \begin{pmatrix} 1 \\ 0 \\ 1 \end{pmatrix}$, so $T\begin{pmatrix} -1 \\ 2 \\ 1 \end{pmatrix} = T(v - 2w) = T(v) - 2T(w) = 4v - 2v = 2v$. We can also use "source-image" table method: We write the given information such that each row on the left is a source vector and its image is written as a row on the right. We do row operations to get new sources and their respective images. This works since the row operations are linear operations. For example: $T(R_3) = T(R_1 - R_2) = T(R_1) - T(R_2)$.

$$\begin{array}{ccc|ccc|l} 1 & 2 & 3 & 4 & 8 & 12 & R_1, \text{ given} \\ 1 & 0 & 1 & 1 & 2 & 3 & R_2, \text{ given} \\ \hline 0 & 2 & 2 & 3 & 6 & 9 & R_3 = R_1 - R_2 \\ -1 & 2 & 1 & 2 & 4 & 6 & R_4 = -R_2 + R_3 \end{array}$$

16). Which of the following is true for $x \in (1, 2)$?

(A) $0 \leq \frac{\log(x)}{x-1} \leq 0.5$

(B) $0.3 \leq \frac{\log(x)}{x-1} \leq 0.8$

(C) $0.4 \leq \frac{\log(x)}{x-1} \leq 0.9$

(D) $0.5 \leq \frac{\log(x)}{x-1} \leq 1$

(E) None of the above.

Answer is D. It is possible to calculate the derivative, to conclude that the function is monotonous decreasing, and evaluate $\log(2)$ using Taylor's expansion. However, a more elegant solution facilitates the Mean Value Theorem, namely for $x \geq 1, \exists c \in (1, x)$ such that $\log'(c) = \frac{1}{c} = \frac{\log(x) - \log(1)}{x-1}$. Since $\log(1) = 0$, restrict x to be in $(1, 2)$ to get a direct evaluation for $\frac{\log(x)}{x-1}$ on the interval. This proves that at least D is true. Then, notice that the limit of the function near $x = 1$ is 1, so the function admits values infinitely close to 1. This eliminates all the rest of the answers.

17). Let $(G, *)$ be a finite group, with identity e. Define a new binary operation on G denoted by $\otimes : G \times G \to G$, such that for all $a, b, c, d \in G$, $(a * b) \otimes (c * d) = (a \otimes b) * (c \otimes d)$, and exists $e_\otimes \in G$ such that for all $a \in G$, $a \otimes e_\otimes = e_\otimes \otimes a = a$. Which of the following is true?

I $\forall a, b \in G, a \otimes b = a * b$.

II $e = e_\otimes$.

III G is abelian.

(A) I

(B) II and III

(C) I and II

(D) I and III

(E) I, II and III

Answer is C. Observe the following calculation:

$e = e*e = (e \otimes e_\otimes) * (e \otimes e_\otimes) = (e * e_\otimes) \otimes (e * e_\otimes) = e_\otimes \otimes e_\otimes = e_\otimes$. Now, using $e = e_\otimes$: $a*b = (a \otimes e) * (b \otimes e) = (a*e) \otimes (b*e) = a \otimes b$. The latter identifies the "new operation" with the "old operation", and the given becomes $(a*b)*(c*d) = (a*b)*(c*d)$, which is true for each group. Therefore, G may not be abelian.

18). Which of the following is true about a group of order 6?

(A) It is Abelian.

(B) It is cyclic

(C) It has a member of order 2.

(D) It has a member of order 4.

(E) None of the above.

Answer is C. There are 2 groups of order 6: The cyclic $\mathbb{Z}_2 \times \mathbb{Z}_3$ and the dihedral group of the triangle, denoted D_6 (or sometimes D_3). Both are well known and both has an element of order 2. D_6 is neither cyclic nor abelian - eliminate A and B, By Lagrange's theorem, the order of a member must divide the order of the group - eliminate D. Another solution: Using the Sylow theorems, since $6 = 2 \cdot 3$, it has a 2-Sylow subgroup which is a cyclic subgroup generated by an element of order 2.

19).
$$\lim_{n \to \infty} \inf_{x \in \mathbb{R}} (e^x - nx) = ?$$

(A) $+\infty$

(B) $-\infty$

(C) e

(D) $\frac{1}{e}$

(E) Does not exists.

Answer is B. First, take a derivative of the function $f(x) = e^x - nx$. $f'(x) = e^x - n$. By sketching the graph for $f'(x)$, one can see that the derivative changes from being negative to positive, creating a minimum point at $x = \log(n)$. Note that this is a global minimum since $\lim_{x \to \pm\infty}(e^x - nx) = +\infty$. So $\inf_{x \in \mathbb{R}} e^x - nx = \min_{x \in \mathbb{R}} e^x - nx = n - n\log(n) = n(1 - \log(n))$. The limit of the latter is of the form $+\infty \cdot -\infty = -\infty$.

20). Given $\forall x \in \mathbb{R}$, $f(x) = f(x^2)$, what is true about $f(x)$?

(A) $f(x)$ is bounded.

(B) If $f(x)$ is a polynomial, $f(x)$ has a bounded derivative.

(C) If $f(x)$ is a polynomial, $f(x) = 0$.

(D) $f(x)$ is odd.

(E) Such $f(x)$ does not exists.

Answer is B. Say $f(x)$ is a polynomial. Then $f(x) - f(2)$ is a polynomial with infinitely many roots (for example, at $x = 2, 4, 16, ...$), thus $f(x) - f(2) = 0$ which implies $f(x)$ is constant (not necessarily zero), let alone with bounded derivative. Notice that $f(x)$ must be even, so it is not odd and we thus eliminate D.

To eliminate A, one has to rely on intuition that building a unbounded function is possible. Here is a full construction:

First, we already established that such $f(x)$ must be even, so it is sufficient to define $f(x)$ on the positive real numbers. Next, we define an equivalence relation: $x \sim y$ if there exists an integer z such that $x^{(2^z)} = y$. (Easy to see that this is an equivalence relation). By definition, $f(x)$ must be constant on the equivalence classes. For each class $S(x)$ define $f(x) = f(S(x))$ to be the minimum natural number in $S(x)$, or 0 otherwise. Now $f(x)$ is well defined. To see that $f(x)$ is unbounded, we just need to observe that there are infinitely many classes that contain natural numbers. Just look at the prime numbers – each must be in a different class.

21).
$$\int_0^{\frac{\pi}{2}} |\cos(x)| + \int_{\frac{\pi}{2}}^{\pi} |\sin(x)| + \int_{\pi}^{\frac{3\pi}{2}} |\cos(x)| + \int_{\frac{3\pi}{2}}^{2\pi} |\sin(x)| = ?$$

(A) 1

(B) 2

(C) 3

(D) 4

(E) π

Answer is D. This question tries to throw you off your game with confusing signs. However, recall the following fact: the area under the graph of $\sin(x)$ and $\cos(x)$ is 1 for every quarter period. That fact makes this question trivial.

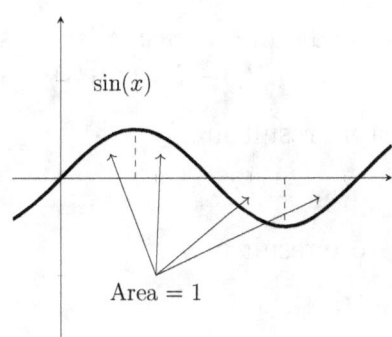

22). Let $f(x) = x^{-a}$, $g(x) = x^{-b}$, $a, b \in (0, 10)$. Which of the following is true?

(A) $f(x) > g(x)$ when $x \in (3, 6)$ and $a > b > 1$.

(B) $f(x) < g(x)$ when $x \in (0, 4)$ and $a > b$.

(C) $f(x) > g(x)$ when $x \in (0.5, 0.9)$ and $a < b$.

(D) $f(x) < g(x)$ when $x \in (0.3, 0.5)$ and $a < b < 1$.

(E) None of the above.

Answer is D. It does not matter if a, b are more or less then 1. The relation is determined by which of the exponent is bigger, and the interval in question. Note that the family of $\{x^{-a}\}$ functions change behavior around 1 - eliminate B. The graphs below are typical for those kind of functions, and only D is correct.

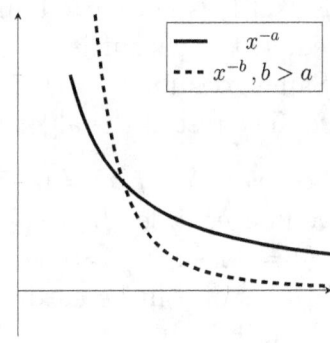

23). Suppose we have an algorithm B that is given as input two **sorted** lists of numbers l_1, l_2 and outputs a **sorted** list of numbers l_3 composed of l_1 and l_2, i.e $l_3 = B(l_1, l_2)$. Consider an algorithm A for sorting a list of numbers, when the an input is a single list of numbers l(not necessarily sorted), and the output $\tilde{l}$ is a sorted list of numbers composed of the numbers in l, i.e. $\tilde{l} = A(l)$. The following is pseudo-code describing A:

1. get input l.

2. if l is sorted, output l.

3. else,

3.1 separate l in the middle into two smaller lists, the left part l_1 and the right part l_2, where $0 \leq length(l_1) - length(l_2) \leq 1$.

3.2 calculate $A(l_1)$, store result in x

3.3 $l_3 = x$

3.4 calculate $A(l_2)$, store result in y

3.5 $l_4 = y$

3.6 output $B(l_3, l_4)$

Note that a list of length 1 is sorted. For an input $l = [3, 1, 2, 4]$, what would be the content of l_3 for each time the algorithm A executes step 3.3.

(A) $l_3 = [1, 3]$

(B) $l_3 = [3, 1]$

(C) $l_3 = [1, 3]$, $l_3 = [1]$

(D) $l_3 = [1, 3]$, $l_3 = [3]$

(E) $l_3 = [3]$, $l_3 = [1, 3]$

Answer is E. Follow this recursive algorithm - [3,1,2,4] is not sorted so is divided into $l_1 = [3, 1], l_2 = [2, 4]$. Note that now A is executed on [3,1], **before** any value is stored as l_3 of this phase. [3,1] is not sorted and is divided to [3],[1]. Now A is executed on [3]. It is sorted, so the output is [3]. This is actually the first time we encounter a value of l_3. Same goes for $l_4 = [1]$. Due to step 3.6, we output [1,3] back to l_3 of the first phase. The rest of the algorithm outputs a sorted list [1,2,3,4]

24). For a real function $f(x)$, Let $A_f = \{x \mid f(x) \neq 0\}$ be a subset of $\mathbb{R}$ on which $f(x)$ is non zero. For each positive integer n, let $B_n = \{\frac{m}{n} \mid m \in \mathbb{Z}, \} \cap [0, 1]$, Define a map G on real functions by $G(f) = \limsup_{n \to \infty} \#(A_f \cap B_n)$, where $\#(C)$ is the number of elements in the set C if C is of a finite cardinality, and $\#(C) = \infty$ otherwise. Which of the following is FALSE?

(A) Exists $f(x)$ such that $G(f) = 0$.

(B) Exists a non constant $f(x)$ such that $G(f) = 0$.

(C) Exists a continuous $f(x)$ such that $G(f) = 0$.

(D) Exists $f(x)$ such that the series $\#(A_f \cap B_n)$ does not have a limit.

(E) Exists a prime integer p, such that for all $f(x)$, $G(f) \neq p$.

Answer is E. Notice that B_n contains only rational numbers, so the famous Dirichlet (complement) function, which is 1 on the irrationals and 0 elsewhere, gives an example for A and B since the sequence $\#(A_f \cap B_n)$ would be constant zero.

$f(x) = 0$ is an example for C. For D and E, consider a function which is non zero on the exactly p rationals: $\frac{1}{p}, \frac{2}{p}, ..., \frac{p}{p}$, where p is prime. For $n = kp$ we have $\#(A_f \cap B_n) = p$, and 0 otherwise. This sequence does not have a limit (eliminate D), but $\limsup \#(A_f \cap B_n)$ is p - E is FALSE.

25). Let a, b be integers such that $5a + 9b$ is divisible by 17. Which of the following is not guaranteed to be divisible by 17?

(A) $2a + 2b$

(B) $8a + 11b$

(C) $11a + 13b$

(D) $13a + 3b$

(E) $a + 12b$

Answer is A. For example take $a = -2, b = 3$ and plug in all the options. Another way to see it is to add $5a + 9b$ to itself multiple times to get more integers divisible by 17, and keep track of the coefficients modulo 17: $(5, 9) \to (10, 1) \to (15, 10) \to (3, 2) \to (\mathbf{8, 11}) \to (\mathbf{13, 3}) \to (\mathbf{1, 12}) \to (6, 4) \to (\mathbf{11, 13})$, all of them are divisible by 17, which eliminates all answers but (A).

26). Let $f : (1, \infty) \to [0, \infty)$ be a function, such that the improper integral $\int_1^\infty f(x)\, dx$ converges. Which of the followings true.

    I $\lim_{x \to \infty} f(x)$ exists.

    II If $f(x)$ is monotonous decreasing, then $\lim_{x \to \infty} f(x)$ exists.

    III If $f(x)$ has derivatives of every order then $\lim_{x \to \infty} f(x)$ exists.

(A) I

(B) II

(C) III

(D) II and III

(E) I, II and III

Answer is B. Both I and III can be eliminated by the famous example of a function composed of rectangles of base of length $\frac{1}{n^3}$ and a height of n. Such a function is unbounded, the improper integral converges, however $\lim_{x \to \infty} f(x)$ does not exists. This function can be made C^∞ (i.e. has derivatives of any kind). II is correct since if f is monotonous decreasing, the limit must exists since f is bounded below.

27). Let $A = \{(x, \sin(\frac{1}{x})) \mid x \in (0, 1)\} \cup \{0\} \times [0, 1]$ be a subset of $\mathbb{R}^2$. Which of the following is true about A?

I A is path connected.

II A is connected.

III A is closed.

(A) I, II

(B) II

(C) III

(D) II and III

(E) I, II and III

Answer is B. No one expects you to give a proof. We just need to recall that A is (some version of) a well known example of a connected space which is not path connected. Note that A is not closed since $(0, -0.5) \notin A$, but the closure of A is $\overline{A} = \{(x, \sin(\frac{1}{x})) \mid x \in (0, 1)\} \cup \{0\} \times [-1, 1]$, making $(0, -0.5)$ a limit point not in A.

28). Let:
$$f(x) = \begin{cases} \arctan(x) & |x| < 1 \\ ax^3 + bx^2 + cx + d & |x| \geq 1 \end{cases},$$
where $\arctan(x) = \tan^{-1}(x)$. For which a, b, c, d does $f(x)$ have continuous second derivative?

(A) $a = -\frac{1}{3}, b = 0, c = 1, d = 0$

(B) $a = 1, b = 0, c = 1, d = -2 + \frac{\pi}{4}$

(C) $a = 0, b = \frac{\pi}{4}, c = 0, d = 0$

(D) $a = -\frac{1}{12}, b = 0, c = \frac{3}{4}, d = 0$

(E) None of the above.

Answer is E. This questions should rise a "red flag" since it requires to reconcile 6 constrains with 4 variables. First, since $\arctan(x)$ is an odd function, $d = b = 0$. Now, calculating the 2nd derivative, we conclude that $a = -\frac{1}{12}$ in order to get $f''(\pm 1) = \mp \frac{1}{2}$. This leaves us only with answer D, but plugging in ± 1 we get rational numbers and not $\pm \frac{\pi}{4}$ as desired.

29). Consider the hyperbola $\frac{x^2}{a^2} - \frac{y^2}{b^2} = 1$. Let $(\pm c, 0)$ be the foci ($c = \sqrt{a^2 + b^2}$). Let C be the circle centered at the origin going through the foci. Denote A as the area of the circular sector of C bounded by the x-axis and the hyperbola's asymptote in the first quadrant. For which hyperbola, $A = \pi$?

(A) $\frac{x^2}{9} - \frac{y^2}{9} = 1$

(B) $\frac{x^2}{3} - \frac{y^2}{3} = 1$

(C) $\frac{x^2}{9} - \frac{y^2}{3} = 1$

(D) $\frac{x^2}{3} - \frac{y^2}{9} = 1$

(E) None of the above.

Answer is C. First note that the formula for c is given in the question although it is a required knowledge for the test. This gives us an equation for C: $x^2 + y^2 = a^2 + b^2$. The asymptote is $y = \frac{b}{a}x \Rightarrow \theta = \arctan(\frac{b}{a})$. (The intersection point with C is (a,b), although it is not needed.) So, $A = \frac{\pi r^2}{2\pi/\theta} = \frac{\theta r^2}{2}$. Now we need to plugin and use known values for $\arctan(x)$.

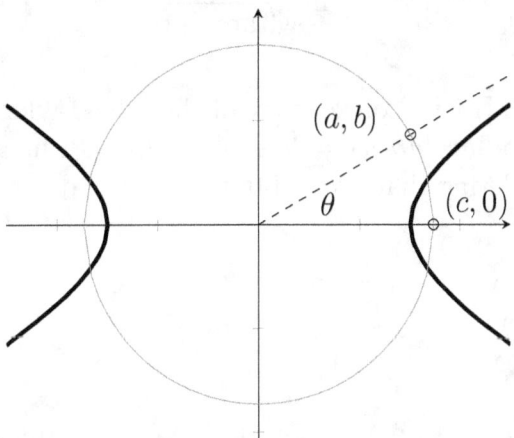

30). Let:
$$A = \int_1^e \frac{\log(x)}{x}\, dx.$$

Which of the following is true?

(A) $0 \leq A < 0.2$

(B) $0.2 \leq A < 0.5$

(C) $0.5 \leq A < 0.7$

(D) $0.7 \leq A < 1$

(E) None of the above.

Answer is C.
$$A = \int_1^e \frac{\log(x)}{x}\, dx = \int_1^e \log(x) \cdot (\log(x))'\, dx = \left.\frac{\log^2(x)}{2}\right|_1^e = \frac{1}{2}.$$

31).
$$\lim_{x \to 1} \frac{(1-x)^3 + 3x - 3}{3\arctan(x-1)} = ?$$

(A) 0

(B) 1

(C) $\frac{1}{3}$

(D) ∞

(E) None of the above.

Answer is B. First, simplify to:
$$\lim_{a \to 0} \frac{3a - a^3}{3\arctan(a)} = ?$$

Now, we can use L'Hopital once or recall that the Taylor expansion of $\arctan(x)$ is $x - \frac{x^3}{3} + O(x^5)$, so $3arctan(a) = 3a - a^3 + O(a^5)$. Plugging the latter in gives us the answer, Considering that around 0, the limit is determined by the lowest power of a.

32). Let $M_3(\mathbb{R})$ be the vector space of 3×3 matrices over $\mathbb{R}$. Let V be a subspace of $M_3(\mathbb{R})$, spanned by:

$$\begin{pmatrix} 1 & 0 & 0 \\ 0 & 1 & 0 \\ 0 & 0 & 1 \end{pmatrix}, \begin{pmatrix} 1 & 1 & 0 \\ 1 & 1 & 0 \\ 0 & 0 & 1 \end{pmatrix}, \begin{pmatrix} 2 & 1 & 0 \\ 0 & 2 & 0 \\ 0 & 0 & 2 \end{pmatrix}, \begin{pmatrix} 2 & 0 & 0 \\ 1 & 2 & 0 \\ 0 & 0 & 2 \end{pmatrix}, \begin{pmatrix} -1 & 5 & 0 \\ 3 & -1 & 0 \\ 0 & 0 & -1 \end{pmatrix}$$

Which of the following linear transformation, V can be the kernel of? (Note: for a field $\mathbb{F}$ denote $\mathbb{F}^n$ as the n dimensional vector space of $n \times 1$ column vectors over $\mathbb{F}$, and denote $\mathbb{F}_n[x]$ to be the vector space spanned by the polynomials of degree $\leq n$ with coefficients in $\mathbb{F}$.

(A) $T: M_3(\mathbb{R}) \to \mathbb{R}_5[x]$

(B) $T: M_3(\mathbb{R}) \to \mathbb{R}$

(C) $T: M_3(\mathbb{R}) \to \mathbb{R}^4$

(D) $T: M_3(\mathbb{R}) \to W$, where W is a subspace of $M_3(\mathbb{R})$ spanned by all the diagonal matrices.

(E) None of the above.

Answer is A. First notice that V is a subspace of dimension 3: use the first matrix to "eliminate" the diagonal of each matrix, i.e.:

$$V = \text{span}\left\{\begin{pmatrix} 1 & 0 & 0 \\ 0 & 1 & 0 \\ 0 & 0 & 1 \end{pmatrix}, \begin{pmatrix} 0 & 1 & 0 \\ 1 & 0 & 0 \\ 0 & 0 & 0 \end{pmatrix}, \begin{pmatrix} 0 & 1 & 0 \\ 0 & 0 & 0 \\ 0 & 0 & 0 \end{pmatrix}, \begin{pmatrix} 0 & 0 & 0 \\ 1 & 0 & 0 \\ 0 & 0 & 0 \end{pmatrix}, \begin{pmatrix} 0 & 5 & 0 \\ 3 & 0 & 0 \\ 0 & 0 & 0 \end{pmatrix}\right\}.$$

Now it is clear that the first, third and forth matrix is a basis for V. Next, recall that any linear transformation $T: U \to W$ satisfy $\dim \ker T + \dim \operatorname{Im} T = \dim U \Rightarrow \dim U - \dim \ker T = \dim \operatorname{Im} T \leq \dim W$. So, in our case, $\dim W \geq 9 - 3 = 6$. Only answer A presents a target vector space of dimension 6 or more. Note that by definition, $\mathbb{R}_5[x]$ is of dimension 6. Also note, that when the dimensions agree, such a transformation always exists.

33). Observe the following graphs of $\arctan(x), \log(x)$ and $\frac{1}{x}$:

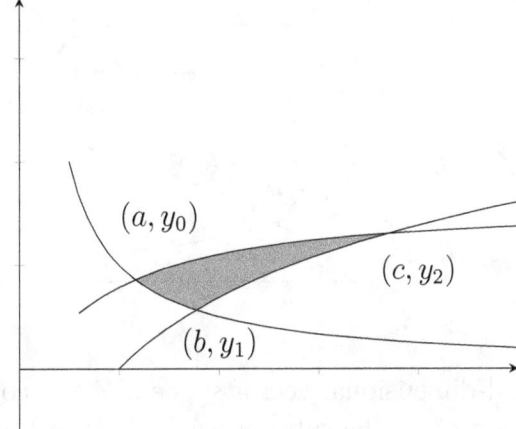

$(a, y_0), (b, y_1), (c, y_2)$ are the coordinates of the intersection points as seen in the figure above. The shaded region is calculated by the following integral($a \leq b \leq c$):

$$\int_a^b \int_{f(x)}^{g(x)} dy\, dx + \int_b^c \int_{h(x)}^{q(x)} dy\, dx$$

Which of the following is true?

I $a \cdot \arctan(a) = b \cdot \log(b)$

II $h(x)$ may be equal to $\log(x)$

III $g(x)$ may be equal to $q(x)$

(A) I

(B) II

(C) III

(D) I and II

(E) I, II and III

Answer is E. First we need to identify the intersection points and the graphs, based on the known behaviour of the functions ($\frac{1}{x}$ decreasing, $\log(x)$ increasing to ∞ and $\log(1) = 0$, $\arctan(x)$ bounded). See figure below. This makes the integral take the following form:

$$\int_a^b \int_{\frac{1}{x}}^{\arctan(x)} dy\, dx + \int_b^c \int_{\log(x)}^{\arctan(x)} dy\, dx$$

The intersection points must satisfy: $\frac{1}{a} = \arctan(a)$, $\frac{1}{b} = \log(b)$, $\log(c) = \arctan(c)$. This shows that I, II and III are all true statements.

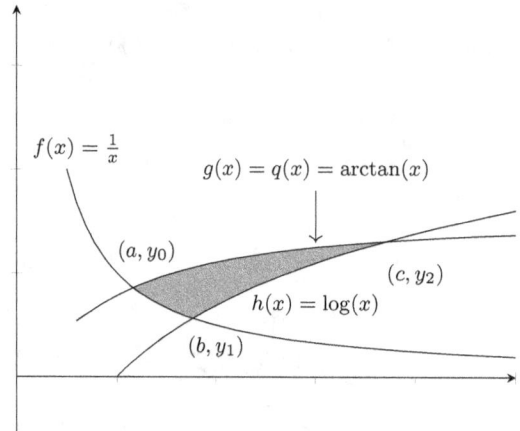

34). Let $\mathbb{C}^4$ denote the 4-dimensional vector space of 4×1 column vectors over $\mathbb{C}$. Let V be the intersection of all the subspaces of $\mathbb{C}^4$. Which of the following is true?

I If V is a vector space, then $\dim V = 0$.

II The cardinality of V is 1.

III If V is a vector space, then the empty set $\emptyset$ is a basis of V.

(A) I

(B) II

(C) III

(D) I and II

(E) I, II and III

Answer is E. This question is based on well-known misconceptions about the zero vector space. The intersection of all subspaces is the zero subspace, so $V = \{0\} =$

$span(\emptyset)$. The empty set is the only basis of V, the dimension is (the number of element in a basis =) 0, and V contains only one element - the zero vector.

35). Observe the following matrix:

$$\begin{pmatrix} i & i & 0 & i & 0 & i \\ 0 & 1 & -1 & -1 & 2 & 2 \\ 1 & 0 & 1 & 2 & -2 & -1 \end{pmatrix}$$

What is its row vectors space dimension?

(A) 2

(B) 3

(C) 4

(D) 5

(E) 6

Answer is A. Recall that the dimension of the row space is equal to the dimension of the column space (both equal to the rank). Immediately, we can eliminate C, D, E since the dimension of the column space is bounded by 3. It is easier to do "column" reduce and to conclude that the dimension is 2. Another method is to divide the first row by i and observe that it is the sum of the other 2 rows.

36). Let E be the part of the ellipse $x^2 + \frac{y^2}{4} = 1$ in the first quadrant. Let A be a rectangle with a vertex at the origin and an opposing vertex on E. What is the maximum possible value of the area of such A?

(A) $\frac{1}{2}$

(B) 1

(C) $\sqrt{2}$

(D) $\frac{\sqrt{2}}{2}$

(E) None of the above

Answer is B. We can formulate the problem by: maximize xy under the constrain $x^2 + \frac{y^2}{4} - 1 = 0$. Solve by the usual method of Lagrange multipliers (useful to notice that A would have maximal area when both $x \neq 0, y \neq 0$, and both x, y are positive):

$$F(x,y,\lambda) = xy + \lambda(x^2 + \frac{y^2}{4} - 1)$$

$$F_x = y + 2\lambda x = 0 \Rightarrow 2\lambda = -\frac{y}{x}$$

$$F_y = x + 2\lambda \frac{y}{4} = 0 \Rightarrow x - \frac{y^2}{4x} = 0 \Rightarrow 4x^2 = y^2 \Rightarrow 2x = y$$

$$F_\lambda = 0 \Rightarrow x^2 + \frac{(2x)^2}{4} = 1 \Rightarrow x^2 = \frac{1}{2} \Rightarrow y^2 = 2 \Rightarrow (xy)^2 = 1 \Rightarrow xy = 1.$$

37). Let E be the graph of $y = \frac{1}{\sqrt{x}}$. Let A be a rectangle with a vertex at the origin and an opposing vertex on E. What is the maximum possible area of such a rectangle A?

(A) $\frac{1}{2}$

(B) 1

(C) $\sqrt{2}$

(D) $\frac{\sqrt{2}}{2}$

(E) None of the above

Answer is E. You can solve using Lagrange multipliers and get that there are no extrema points, or you can observe that $xy = \frac{x}{\sqrt{x}} = \sqrt{x}$. So the area increases without a bound as $x \to \infty$.

38). Consider the following argument:

a) Let the real sequences $a_n \to L$ and $b_n \to L$, as $n \to \infty$. Then,

b) $a_n - b_n \to 0$, as $n \to \infty$. Then,

c) For all $\epsilon > 0$, exists a sub-sequence n_k such that $\sum_{k=1}^{\infty} |a_{n_k} - b_{n_k}| < \epsilon$. Then

d) Exists a sub-sequence n_k such that $a_{n_k} = b_{n_k}$

Which of the following is true?

I a) does not imply b).

II b) does not imply c).

III c) does not imply d).

(A) I

(B) II

(C) III

(D) II and III

(E) I and III

Answer is C. a) implies b) by arithmetic of sequences. b) implies c) since for every $c_n \to 0$ exists a sub-sequence such that $|c_{n_k}| < \frac{\epsilon}{2^k}$, and observe that $\sum_{k=1}^{\infty} |c_{n_k}| < \epsilon \sum_{k=1}^{\infty} \frac{1}{2^k} = \epsilon$. However, the last implication is false. For example, consider the case $a_n = \frac{1}{n}, b_n = -\frac{1}{n}$.

39). Let A be the following matrix:
$$\begin{pmatrix} 1 & 2 \\ 2 & 1 \end{pmatrix}$$

Which of the following is A^5?

(A) $\begin{pmatrix} 1 & 2 \\ 2 & 1 \end{pmatrix}$

(B) $\begin{pmatrix} 121 & 122 \\ 122 & 121 \end{pmatrix}$

(C) $\begin{pmatrix} 121 & 2 \\ 2 & 121 \end{pmatrix}$

(D) $\begin{pmatrix} 212 & 121 \\ 121 & 212 \end{pmatrix}$

(E) $\begin{pmatrix} 112 & 221 \\ 221 & 130 \end{pmatrix}$

Answer is B. The smart way to solve this problem is by following these steps. First compute the eigenvalues. 3 is a eigenvalue with eigenvector $\begin{pmatrix} 1 \\ 1 \end{pmatrix}$ (shortcut: the sum of entries in each row is 3). The trace is 2, so the other eigenvalue is -1 (sum of eigenvalues is the trace) and its eigenvector must be orthogonal to $\begin{pmatrix} 1 \\ 1 \end{pmatrix}$, since this is a symmetric matrix. Let us use $\begin{pmatrix} 1 \\ -1 \end{pmatrix}$. The eigenvalues of A^5 would be 3^5 and -1^5 so we expect A^5 to have a trace of $3^5 - 1 = 9 \cdot 9 \cdot 3 - 1 = 81 \cdot 3 - 1 = 243 - 1 = 242$.

This eliminates A and D. We also expect determinant of $-1 \cdot 3^5 < 0$, so eliminate C. Now we can calculate the determinant or notice that E is not symmetric, but in our case:

Denote $P = \begin{pmatrix} 1 & 1 \\ 1 & -1 \end{pmatrix}$,

By using a formula: $P^{-1} = -0.5 \begin{pmatrix} -1 & -1 \\ -1 & 1 \end{pmatrix} = 0.5P$,

So $P^{-1}AP = \begin{pmatrix} 3 & 0 \\ 0 & -1 \end{pmatrix} \Rightarrow P^{-1}A^5P = \begin{pmatrix} 3^5 & 0 \\ 0 & -1 \end{pmatrix} \Rightarrow A^5 = P \begin{pmatrix} 0.5 \cdot 3^5 & 0 \\ 0 & -0.5 \end{pmatrix} P$,

Denote $A^5 = B$, and observe: $B^{tr} = B$, eliminating E.

Another (advanced) consideration that we can use is the fact that $\begin{pmatrix} 1 \\ 1 \end{pmatrix}$ is an eigenvector of A^5 corresponding to $3^5 = 243$. This eliminates all the answers except for B.

40). Let $f(x) = \int_1^{x^2} \frac{sin(x)}{t} dt$.
$$\lim_{x \to 1} \frac{f(x)}{x-1} = ?$$

(A) 0

(B) 1

(C) $\sin(1)$

(D) $2\sin(1)$.

(E) None of the above.

Answer is D. Either notice that $f(x) = \log(x^2)\sin(x)$ and use L'Hopital, or take the derivative of $f(x)$ using Leibniz integral rule, as follows: $\lim_{x \to 1} \frac{f(x)}{x-1} = \lim_{x \to 1} \frac{f'(x)}{1} = \int_1^{x^2} \frac{cos(x)}{t} dt + 2x \cdot \frac{sin(x)}{x^2} = 0 + 2\sin(1) = 2\sin(1)$

41). Let $f(x,y) = 0$ be a solution of the differential equation $3y(1+x)\,dx + 3x\,dy = 0$ such that $y = 1$ when $x = 1$. What is y when $x = 2$?

(A) 0

(B) 1

(C) e

(D) $6e^2$

(E) None of the above.

Answer is E. We solve this using the regular methods:

Denote $P = 3y(1+x), Q = 3x$.
Observe $P_y = 3(1+x), Q_x = 3$, so the equation is not exact.
$P_y - Q_x = 3x \Rightarrow \dfrac{P_y - Q_x}{Q} = 1$, so we can use an integrating factor of $e^{\int 1\,dx} = e^x$
Now the following is exact: $3y(1+x)e^x\,dx + 3xe^x\,dy = 0 \Rightarrow F(x,y) = 3xye^x + C = 0$.

Use the given $F(1,1) = 0$ to deduce $C = -3e$, So $0 = F(2,y) = 6e^2 y - 3e \Rightarrow y = 0.5e^{-1}$. Thus the answer is E. We can get the same answer if we solve as a separable equation.

42). The following represents a portion of the graph of the **derivative** of $f(x)$, on the interval (a, b):

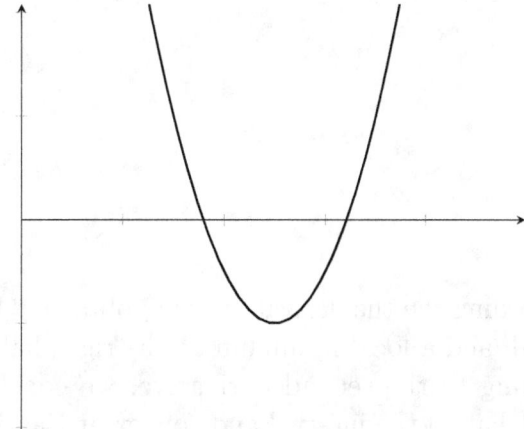

Which of the following may represent the graph of $f(x)$ on the interval (a, b)?

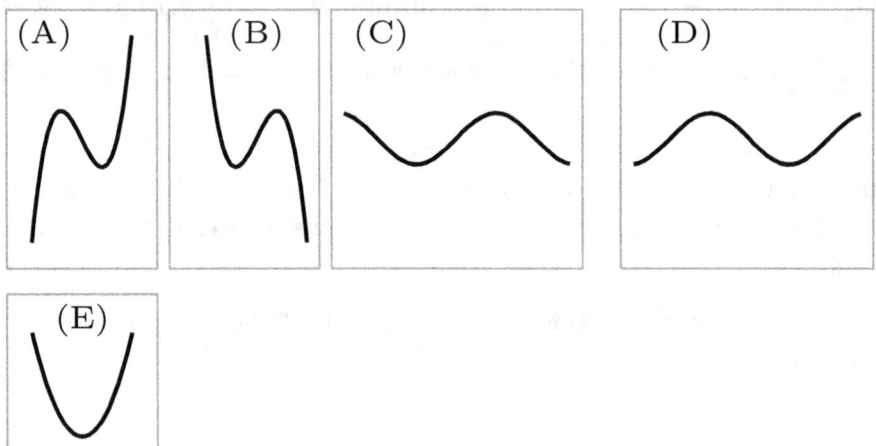

Answer is A. According the the derivative, $f(x)$ obtains a local maximum at the left half of the interval, and a local minimum at the right half. That eliminates B,C,E. Moreover, according to the second derivative, we conclude that the function is concave at the left half of the interval and convex at the right half. That eliminates D (in fact, that eliminates all the answers expect A).

43). Which of the following real matrices, has i and $1+i$ as eigenvalues?

(A) $A = \begin{pmatrix} 8 & 0 & 4 & 8 \\ 4 & -12 & -8 & 0 \\ -2 & 20 & 14 & 4 \\ -3 & -8 & -7 & -2 \end{pmatrix} \cdot \begin{pmatrix} \frac{1}{4} & 0 & 0 & 0 \\ 0 & \frac{1}{4} & 0 & 0 \\ 0 & 0 & \frac{1}{4} & 0 \\ 0 & 0 & 0 & \frac{1}{4} \end{pmatrix}$

(B) $B = \begin{pmatrix} 1 & 4 & -1 & 0 \\ 2 & 8 & -2 & 0 \\ 5 & -4 & 3 & 0 \\ 5 & 0 & -5 & -4 \end{pmatrix} \cdot \begin{pmatrix} \frac{1}{4} & 0 & 0 & 0 \\ 0 & \frac{1}{4} & 0 & 0 \\ 0 & 0 & \frac{1}{4} & 0 \\ 0 & 0 & 0 & \frac{1}{4} \end{pmatrix}$

(C) $C = \begin{pmatrix} 1 & 4 & -1 & 0 \\ -5 & 9 & -1 & 0 \\ 5 & -4 & 3 & 0 \\ 5 & 0 & -5 & -4 \end{pmatrix} \cdot \begin{pmatrix} \frac{1}{4} & 0 & 0 & 0 \\ 0 & \frac{1}{4} & 0 & 0 \\ 0 & 0 & \frac{1}{4} & 0 \\ 0 & 0 & 0 & \frac{1}{4} \end{pmatrix}$

(D) $D = \begin{pmatrix} 1 & 4 & -1 & 0 \\ -5 & 8 & -1 & 0 \\ 5 & -4 & 3 & 0 \\ 5 & 0 & -5 & -5 \end{pmatrix} \cdot \begin{pmatrix} \frac{1}{4} & 0 & 0 & 0 \\ 0 & \frac{1}{4} & 0 & 0 \\ 0 & 0 & \frac{1}{4} & 0 \\ 0 & 0 & 0 & \frac{1}{4} \end{pmatrix}$

(E) $E = \begin{pmatrix} 1 & 4 & -1 & 0 \\ -5 & 8 & -1 & 0 \\ 5 & -4 & 4 & 0 \\ 5 & 0 & -5 & -4 \end{pmatrix} \cdot \begin{pmatrix} \frac{1}{4} & 0 & 0 & 0 \\ 0 & \frac{1}{4} & 0 & 0 \\ 0 & 0 & \frac{1}{4} & 0 \\ 0 & 0 & 0 & \frac{1}{4} \end{pmatrix}$

Answer is A. Since the matrix is real, its characteristic polynomial is real. if $i, 1+i$ are its roots, so is $-i$ and $1-i$. So the trace must be (the sum of the eigenvalues, so) 2. Note, that by observing only the left matrices, the trace should be 8. This eliminates C, D, E. For B, notice that the first and second row are linearly dependent, so the matrix has a zero determinant, although the product of the eigenvalues is non-zero. Therefore, A must be the correct answer.

44). Let $M_3(\mathbb{R})$ be the vector space of 3×3 matrices over $\mathbb{R}$. Which of the following is true?

I Exists a basis of $M_3(\mathbb{R})$ which contains no matrices with zero trace.

II Exists a basis of $M_3(\mathbb{R})$ which contains exactly 8 matrix with zero trace.

III Exists a basis of $M_3(\mathbb{R})$ which contains exactly 9 matrix with zero trace

(A) I

(B) II

(C) I,II

(D) II,III

(E) I, II and III

Answer is C. First recall that $M_3(\mathbb{R})$ is a vector space over $\mathbb{R}$ with matrices addition and scalar multiplication. The latter implies that the function $trace : M_3(\mathbb{R}) \to \mathbb{R}$ is a linear transformation. It is certainly surjective, so the kernel (matrices with zero trace) form a subspace of dimension 8. This makes II true, and III false, since there are no 9 matrices of zero trace that are linearly independent. Now, take a basis of ker($trace$), $\{v_1, ..., v_8\}$ and complete it to a basis of $M_3(\mathbb{R})$: $\{v_1, ..., v_8, w\}$. Note that $tr(w) \neq 0$. Now observe the the following set $\{v_1 + w, v_2 + w, ..., v_8 + w, w\}$ is a basis for $M_3(\mathbb{R})$ and each of non zero trace.

45). Let $f(z) = \frac{1}{z^2 - a^2}$ be a complex function with a real parameter $a \in \mathbb{R}$. Let C be the circle $|z| < R$, $R \in \mathbb{R}$, $R > 0$ oriented counterclockwise. For which choice of R and a is the integral $\oint_C f(z)\, dz$ non-zero (as long as C does not intersect $(\pm a, o)$)?

(A) Any R and a, as long as C does not intersect $(\pm a, o)$.

(B) Any R, $a = 0$.

(C) $a > 0, R > a$.

(D) $a > 0, R < a$

(E) Such choice is not possible.

Answer is E. If $a = 0$, $f(z) = \frac{1}{z^2}$, a function with anti derivative, so any integral on a closed curve would be zero (also. 0 is a pole with residue 0). For $a > 0$, $f(z) = \frac{1}{z-a} \cdot \frac{1}{z+a}$. C may include both poles or neither. In the latter, the integral is zero. In the former, the integral is $2\pi i \cdot (Res(a, f(z)) + Res(-a, f(z))) = 2\pi i \cdot (\frac{1}{2a} + \frac{1}{-2a}) = 0$.

46). Let $f(z) = \frac{1}{z(z-1)(z-2)}$ be a complex function. Denote the principle part of the Laurent series of $f(z)$ in $\{|z| > 2\}$ as $\sum_{n=1}^{\infty} c_n z^{-n}$. Denote $\tilde{n}$ as the smallest n such that $c_n \neq 0$. What is the value of $c_{\tilde{n}}$?

(A) 1

(B) i

(C) 3

(D) $3i$

(E) None of the above.

Answer is A. First, we need to decompose to partial fractions: $f(x) = \frac{0.5}{z} + \frac{-1}{z-1} + \frac{0.5}{z-2}$. The first term is already in its "Laurent" form. For the other terms: $\frac{1}{z-a} = \frac{1}{z} \frac{1}{1 - \frac{a}{z}} = \frac{1}{z}\left(1 + \frac{a}{z} + \left(\frac{a}{z}\right)^2 + ...\right)$. Now it is a matter of technical calculation - sum the contribution of each term for each degree n:

$$n = 1: 0.5 - 1(1) + 0.5(1) = 0$$
$$n = 2: 0 - 1(1) + 0.5(2) = 0$$
$$n = 3: 0 - 1(1^2) + 0.5(2^2) = 1$$

Notice, that we can ignore completely from $\frac{1}{z}$ and solve for $\frac{1}{(z-1)(z-2)}$ since $\frac{1}{z}$ only changes $\tilde{n}$ but does not change the answer.

47). A fair roulette outputs a random integer number between 0 and 36 at every spin. The gamblers can bet on ODDS (all the odd numbers) or EVENS (all the even numbers) with any amount of money. In case of a successful gamble, the gambler gets back twice the amount he/she put on stake, otherwise the amount is forfeited. A sophisticated gambler bets according to the following algorithm:

1). Set x to be $1,

2). Bet on ODDS with the amount of x, let the roulette spin.

3). If the output is an odd number, collect a profit of x dollars and stop gambling.

4). Else, set x to be $2x$ (double the value of x) and repeat algorithm from 2).

Which of the following statements is true?

(A) There is a 0.05% chance that the algorithm will never stop.

(B) When the algorithm stops, the gambler, in total, will lose money with probability ≥ 0.5.

(C) There is more that 50% chance that the gambler would lose at least 2^{32} at some point of the algorithm

(D) When the algorithm stops, the gambler, in total, makes a profit of at least $1 with probability ≥ 0.8.

(E) None of the above.

Answer is D. Notice that the algorithm continues as long as the roulette outputs an even number, and the amount of money the gambler loses is: $1 + 2 + 4 + 8 + ...$ So the gambler loses a total of $2^n - 1$ after n unsuccessful bets. However, when the $n+1$ bet is successful, he/she wins 2^n dollars, which leaves him/her with a total profit of exactly $1. Notice that the probability of the number of bets made bigger then k is (as a geometric random variable) $p(EVENS)^k$, so as long as $p(EVENS) < 1$ (and it is, since the roulette is fair), $p(\text{num of bets} > k) \to 0$. This proves that the algorithm will stop with probability 1, and the gambler will win in total $1 with probability 1.

48). Let $f(z) = \sum_{n=1}^{\infty}(-1)^{n-1}\frac{1}{4^n}z^{2n-1}$. Denote $A = \{z \,|\, z \text{ is a pole of } f(z)\}$ and $r = \inf_A\{|z|\}$. Let z_0 be a pole of $f(z)$ such that $|z_0| = r$, if such z_0 exists. What is a possible value of z_0?

(A) -4

(B) i

(C) $1+i$

(D) $-2i$

(E) Such z_o does not exists.

Answer is D. First we need to calculate the radius of convergence. We want to use the formula (Cauchy-Hadamard) $\frac{1}{R} = \limsup |a_n|^{\frac{1}{n}}$. A common mistake is to take $a_n = \frac{1}{4^n}$ and to get a radius of 4. We need to notice that the sum is given over x^{2n-1} so in fact, $a_{2n-1} = \frac{1}{4^n} \Rightarrow a_k = \frac{1}{4^{(k+1)/2}} = \frac{1}{2^{(k+1)}}$. (Note that the coefficients in the even spots are 0, but that does not change the radius formula since it is a "lim sup").

So the radius of convergence is actually 2, which means that the function $f(z)$ has a pole of radius 2, and it is the minimal radius among its poles. Only D satisfies the latter.

49). For $b, c \in \mathbb{Z}$, let $A_{b,c} = \{bx + cy \mid x, y \in \mathbb{Z}\}$ be a subset of the ring $\mathbb{Z}$. We say that V is a "generated" subset of $\mathbb{Z}$ if $\exists v \in V$ such that $\forall w \in V, \exists n \in \mathbb{Z}$ for which $w = nv$. For which of the following pairs (b, c) the set $A_{b,c}$ is NOT generated?

(A) $(1, 2)$

(B) $(2, 3)$

(C) $(16, 48)$

(D) $(540, 448)$

(E) None.

Answer is E. $A_{b,c}$ is an ideal of $\mathbb{Z}$ for any choice of b, c. By using the euclidian algorithm (or by just knowing this fact on $\mathbb{Z}$) one can prove that $A_{b,c}$ is generated by $gcd(b, c)$ (In some textbooks, this is even the definition of the gcd). Note that the latter is true even when b or c is 0 or 1; if one of them is 1, we get the entire ring which is generated by 1. If $b = 0$ then $A_{0,c}$ is generated by c.

50). 35 people are attending a party. Any pair of people may or may not talk during the night. At the end of the night, each participant will have conversed with 0,1,2,...,33 or 34 different people. Let p be the probability that there exist 2 different people which have conversed with the same number of people. What is true about p?

(A) $0 \leq p < 0.2$

(B) $0.2 \leq p < 0.4$

(C) $0.4 \leq p < 0.6$

(D) $0.6 \leq p < 0.8$

(E) $0.8 \leq p \leq 1$

Answer is E. For any participant, the possible number of other people which he/she conversed with is 0,1,2,...,33 or 34. These are 35 different numbers and we have to assign a number to each participant. Notice that if one participant talked to 0 people, there is no one who talked to 34, and vice versa. So, in fact, we need to assign 34 numbers to 35 participants. By the pigeonhole principle, p is 1.

51). Which of the following is of finite or countable cardinality?

    I $\mathbb{Q} \times \mathbb{Q} \times \mathbb{Q}$.

    II $\bigcup_{n \in \mathbb{N}}(\mathbb{Z} + \frac{1}{n})$ where $\mathbb{Z} + a = \{x + a \,|\, x \in \mathbb{Z}\}$.

    III All the subsets of $\mathbb{N}$.

(A) I

(B) I,II

(C) II,III

(D) I,III

(E) I,II and III

Answer is B. Recall, countable union of a countable sets is countable, finite cartesian product of countable sets is countable. However, the power set of a countable set is uncountable.

52). Denote $\mathbb{C}_4[x]$ to be the vector space spanned by the polynomials of degree ≤ 4 with coefficients in $\mathbb{C}$. Let $D : \mathbb{C}_4[x] \to \mathbb{C}_4[x]$ be the derivative linear transformation, i.e. $D(f(x)) = \frac{df(x)}{dx}$. What is true about D? (denote tr as the trace, det as the determinant).

    I $\operatorname{tr}(D) > 3$

    II $\operatorname{tr}(D) = \det(D)$

    III The characteristic polynomial of D is $x^5 + x^3 + x$.

(A) I

(B) II

(C) III

(D) II,III

(E) I,II and III

Answer is B. Take a basis $\{v_1 = 1, v_2 = x, v_3 = x^2, v_4 = x^3, v_5 = x^4\}$. Apply D to get $D(v_1) = 0$, $D(v_2) = 1 = v_1$, $D(v_3) = 2x = 2v_2$, $D(v_4) = 3x^2 = 3v_3$, $D(v_5) = 4x^3 = 4v_4$. In that basis, the matrix representing D is:

$$\begin{pmatrix} 0 & 1 & 0 & 0 & 0 \\ 0 & 0 & 2 & 0 & 0 \\ 0 & 0 & 0 & 3 & 0 \\ 0 & 0 & 0 & 0 & 4 \\ 0 & 0 & 0 & 0 & 0 \end{pmatrix}.$$

Now it is easy to see that $\det(D) = \operatorname{tr}(D) = 0$ and the characteristic polynomial is x^5.

53). For every Lebesgue measurable set A denote $\mu(A)$ as the Lebesgue measure of A. Let $f(x)$ be the following function:

$$f(x) = \begin{cases} x & x \in \mathbb{Q} \\ 0 & \text{otherwise} \end{cases}.$$

Which of the following is true?

I $f(x)$ is a Lebesgue measurable function.

II $\int_\mathbb{R} f(x)\, d\mu < \infty$

III $\int_\mathbb{R} f(x)\, d\mu$ is equal to the (improper) Riemann integral of $f(x)$.

(A) II

(B) I,II

(C) I,III

(D) II,III

(E) I,II and III

Answer is B. $f(x)$ is measurable since every pre-image of an interval contains a subset of the rational (which has measure zero) and possibly contains $\mathbb{R} \setminus \mathbb{Q}$ which is measurable as the complement of $\mathbb{Q}$. $f(x)$ is 0 almost everywhere, so the Lebesgue integral of $f(x)$ is 0. However, $f(x)$ is not Riemann-integrable (in the broad sense); for every interval $[a, b]$, upper Darboux sum would be $b \cdot (b - a)$ while the lower Darboux sum would be 0.

54). Let G be the Dihedral group of order 8. Let n to be the number of elements in G of order less than 3. How many groups of order n exists (up to isomorphism)?

(A) 0

(B) 1

(C) 2

(D) 3

(E) 6

Answer is C. In the Dihedral group of order 8 there are 8 elements altogether, generated by the rotation r (of order 4) and reflection t (of order 2), so $G = \langle r, t \,|\, rt = tr^{-1} = tr^3 \rangle$. Notice that all the members of the form $r^a t$ are of order 2 since $r^a t r^a t = r^a t t r^{-a} = e$. So, the only 2 elements of order 4 are r and r^3. Thus $n = 6$. There are only 2 different groups of order 6 up to isomorphism: the cyclic group C_6 and the Dihedral group of order 6.

55). Let $f(x, y) = x^3 + ay^2 + 4xy$, where $a \neq 0$. Which of the following must be true?

(A) $\exists a \in [2, 6]$ such that $f(x, y)$ has no saddle points.

(B) $\exists a \in [-2, 0)$ such that $f(x, y)$ has 2 distinct minimum points.

(C) $\exists a \in [1, 5]$ such that $f(x, y)$ has one maximum point.

(D) $\exists a \in [-2, 6]$ such that $f(x, y)$ has 2 distinct maximum points.

(E) $\exists a \in [3, 7]$ such that $f(x, y)$ has exactly one saddle point.

Answer is E. For a general $a \neq 0$ observe:

$$f_x = 3x^2 + 4y = 0$$
$$f_y = 2ay + 4x = 0$$
$$\Rightarrow 2y = -4x/a \Rightarrow \text{(using } f_x = 0\text{)} 3x^2 - 8x/a = 0 \Rightarrow x = 0 \text{ or } x = 8/(3a)$$
$$f_{xx} = 6x, f_{yy} = 2a, f_{xy} = 4 \Rightarrow \text{(Hessian) } H = 12xa - 16$$

For $x = 0$ we get $H = -16 < 0 \Rightarrow$ a saddle point. For $x = 8/(3a)$, we have $H = 32 - 16 = 16 > 0$ and $f_{xx} = 16/a$. So for positive a we have a minimum point, and for negative a we have a maximum point.

56). Let $f_n(x) : [0, 1] \to \mathbb{R}$ be a sequence of functions. Which of the following must be true? (Note: a statement in true almost everywhere on an interval I means that the set of all $x \in I$ for which the statement is not true, has a Lebesgue measure zero. Also note that the integrals may be improper).

I $\lim_{n \to \infty} \int_0^1 |f_n(x)| dx = 0$ implies $f_n(x) \to 0$ almost everywhere on $[0, 1]$

II $\lim_{n \to \infty} \int_0^1 f_n(x) dx < \infty$ DOES NOT imply that $\exists N$ such that $f_n(x)$ is bounded on $[0, 1]$ for all $n > N$.

III If $f_n(x)$ converges pointwise to a continuous function $f(x)$ then $\exists N$ such that $f_n(x)$ is bounded on $[0, 1]$ for all $n > N$.

(A) I

(B) II

(C) I,II

(D) II,III

(E) I,II and III

Answer is B. A famous counter example for I is (sometimes called the "typewriter function"): $f_n(x) := \mathbb{1}_{[(n-2^k)/2^k, (n-2^k+1)/2^k]}$ when we choose k to satisfy $2^k \leq n < 2^k + 1$. The special property is that for each point $x \in (0, 1)$ the limit $\lim_{n \to \infty} f_n(x)$ does not exists, although the integral approaches to 0. II is true since we can take

$$f_n(x) = \begin{cases} \frac{1}{\sqrt{x}} & x \in (0, 1] \\ 0 & x = 0 \end{cases},$$

for any n. This integral is constant and finite, while the function is not bounded.

III is false by taking:

$$f_n(x) = \begin{cases} \frac{1}{x} & x \in (0, \frac{1}{n}] \\ 0 & \text{otherwise} \end{cases}.$$

$\lim_{n \to \infty} f_n(x) = 0$ for any x so the limit function is the continuous constant 0 function, however, the functions $f_n(x)$ are not bounded.

57). Let L be the approximation of $\log(x)$ for $x = 1.02$ by its Taylor expansion of order 2 around 1. Denote $r = \log(1.02) - L$. Which of the following is true?

(A) $0.0001 \leq r \leq 0.001$

(B) $-0.01 \leq r \leq -0.001$

(C) $0.00001 \leq r \leq 0.0001$

(D) $-0.001 \leq r \leq -0.0001$

(E) $0.000001 \leq r \leq 0.00001$

Answer is E. To make the calculation easier, let us consider the function as $\log(1+x)$ with its taylor expansion of order 2 around $x = 0$, denoted as $T_2(x)$. The formula for the remainder $f(0.02) - T_2(0.02)$ is $\frac{f^{(3)}(c)}{3!}(0.02)^3$ for some $c \in [0, 0.02]$. $f^{(3)}(x) = \frac{2}{(1+x)^3}$ $\Rightarrow$ the reminder must be non negative, so eliminate B and D. Now we have

$$r = \frac{f^{(3)}(c)}{3!}(0.02)^3 = \frac{\frac{2}{(1+c)^3}}{3!}\frac{1}{50^3} \leq \frac{1}{3 \cdot 50^3} = \frac{1}{3 \cdot 125 \cdot 1000} \leq \frac{1}{100 \cdot 1000} \leq \frac{1}{100000} \leq 0.00001.$$

Thus, E is the correct answer.

58). Let $f(z): X \to \mathbb{C}$ be an analytic function, $X \subset \mathbb{C}$ such that $|f(z)|$ is constant on a non-empty open set of the complex plane, contained in X. Which of the following is true about $f(z)$? (Note: $f(z) = f(x+iy)$, $x, y \in \mathbb{R}$, $f_x(z)$ is the partial derivative of $f(z)$ with respect to x, and similar for y).

(A) $f_x(z)$ and $f_y(z)$ are equal for some subset of the complex plane of infinite cardinality.

(B) $f(z)$ must be unbounded.

(C) $f(z)$ must be constant.

(D) Exists a point $z_0 \in \mathbb{C}$ such that the power series expansion of $f(z)$ around z_0 converges to $f(z)$ for any z such that $|z - z_0| < 1$.

(E) Such $f(z)$ does not exists.

Answer is A. Note that all we know is that $f(z)$ is defined on some subset of $\mathbb{C}$, namely X, and for at least some open ball in that X, $|f(z)|$ is constant. We also know that $f(z)$ is analytic in every point it is defined. For example, take $X = A \cup B, A = \{z : |z| < \frac{1}{2}\}, B = \{z : |z-2| < \frac{1}{2}\}$,

and $f(z)$ is 0 on A and 1 on B. This is a counter example for B,C,D and E; it is a bounded function, non constant on X, and the power expansion around any $z_0 \in X$, cannot converge to $f(z)$ beyond a radius of $\frac{1}{2}$ since the function is not defined there. A is true by applying Cauchy-Riemann on the open set in which $|f(z)|$ is constant:

$$f(z) = u(x,y) + iv(x,y)$$
$$u^2 + v^2 = c \Rightarrow 2u_x u + 2v_x v = 0, 2u_y u + 2v_y v = 0$$
$$\Rightarrow \text{By Cauchy-Riemann: } u_x u - u_y v = 0, u_y u + u_x v = 0 \Rightarrow$$
$$\begin{pmatrix} u & -v \\ v & u \end{pmatrix} \begin{pmatrix} u_x \\ u_y \end{pmatrix} = \begin{pmatrix} 0 \\ 0 \end{pmatrix}$$

The determinant of the above matrix is c. If $c = 0$, then the function is constant 0 on this open set (since $|f(z)| = 0$). Otherwise, if $c \neq 0$ the determinant is invertible, and thus $u_x = u_y = 0$ for that open set. In both cases, combining with Cauchy-Riemann, we get that $u_x = u_y = v_x = v_y = 0$ for some open set in $\mathbb{C}$, which must have infinite cardinality.

59). A fair coin is tossed 10000 times. Let $p_{a,b}$ be the probability that the number of heads H satisfy $a \leq H \leq b$. Which of the following a, b pairs would ensure that $0.37 \leq p_{a,b} \leq 0.45$

(A) $a = 5000, b = 7000$

(B) $a = 5050, b = 5100$

(C) $a = 4975, b = 5025$

(D) $a = 4950, b = 5050$

(E) None of the above

Answer is C. H is a binomial random variable $B(p = 0.5, n = 10000)$, and we can approximate it using normal distribution with average $\mu = np = 5000$ and standard deviation of $\sigma = \sqrt{np(1-p)} = \sqrt{2500} = 50$. Now, it is a matter of recalling known values from the gaussian:

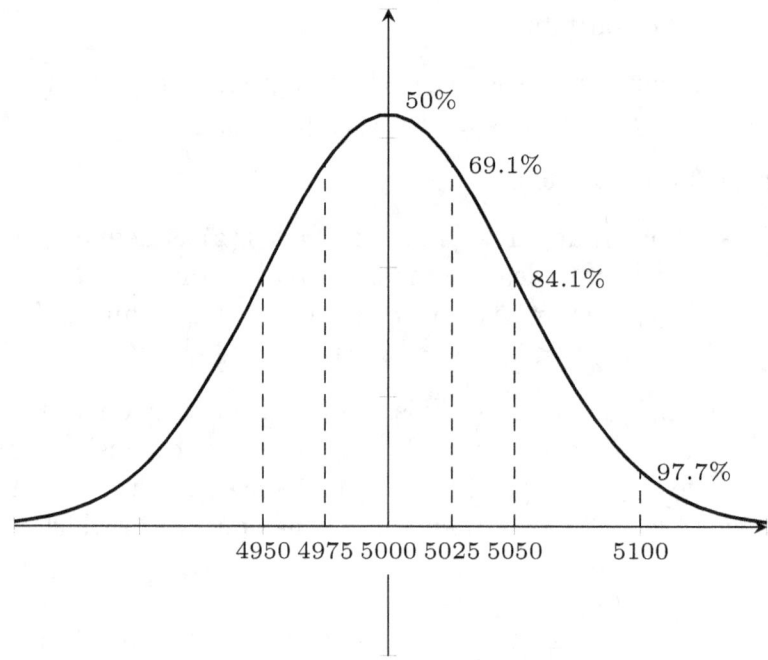

60). The following data points are given: (1,1), (2,3), (4,5), (2,3), (4,5), (5,1). What is the linear regression line equation (minimizing the square distance between the line and the data points)?

(A) $y = \frac{1}{2}x + 1$

(B) $y = \frac{1}{3}x + 1$

(C) $y = \frac{1}{2}x + 2$

(D) $y = \frac{1}{3}x + 2$

(E) $y = \frac{1}{2}x + 2.5$

Answer is D. Well, one question about numerical analysis is bound to show up somewhere. The trick here is to know (or to make an educated guess) that the line must go through the "center of mass" of the data points:

$$\bar{x} = \frac{1+2+4+2+4+5}{6} = 3; \bar{y} = \frac{1+3+5+3+5+1}{6} = 3.$$

The line in D is the only one goes through (3,3).

61). Denote:

$$A = \oint_C ye^{xy}dx + (xe^{xy} + x)dy.$$

What is the value of A, where C is the graph in the following figure, oriented counterclockwise?

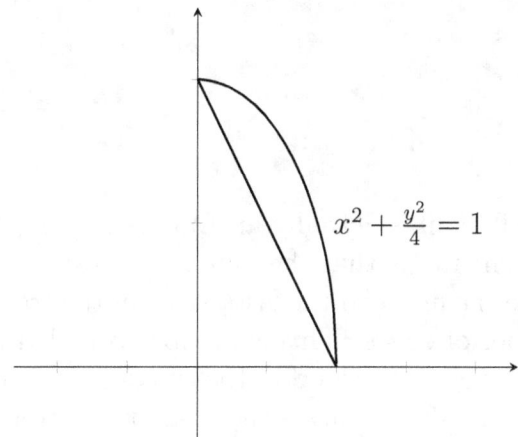

(A) $\pi - 1$

(B) $\pi - 2$

(C) $2\pi - 1$

(D) $2\pi - 2$

(E) None of the above.

Answer is E. Using Green's theorem, we conclude that A can be calculated as a double integral on the region bounded be the graph, where the integrand is:

$$\frac{\partial(xe^{xy} + x)}{\partial x} - \frac{\partial(ye^{xy})}{\partial x} = 1.$$

Thus, A the the area of the region which is the area of a quarter of the ellipse $\pi \cdot 2 \cdot 1/4 = \pi/2$, minus the area of the triangle, which is 1. In total, $A = \frac{\pi}{2} - 1$.

62). A graph is composed of 16 vertices, enumerated $\{v_1, v_2, v_3, v_4, v_5, v_6, v_{2a}, v_{2b}, v_{3a}, v_{3b}, ..., v_{6a}, v_{6b}\}$. 5 edges connecting v_1 to each of $\{v_2, v_3, v_4, v_5, v_6\}$. Each of the vertices $\{v_2, v_3, v_4, v_5, v_6\}$ has an additional 2 edges connecting v_i to v_{ia} and v_{ib}, $2 \le i \le 6$. The graph contains a total of 15 edges.

Define a **Full Trail** to be a finite sequence of vertices and edges, where each edge's endpoints are the preceding and following vertices in the sequence AND the first and the last members of the sequence are vertices AND every vertex which is not the last member of the sequence is followed by an edge AND the sequence contains all the edges in the graph **only once**.

What is the minimum number of edges need to be added to the graph, in order for a full trail to exist.

(A) 0

(B) 1

(C) 6

(D) 7

(E) 8

Answer is D. A Full Trail is actually an Eulerian Trail. A necessary condition for the existence of such trail is that the number of edges connected to each vertex is even, except maybe the first and the last vertex of the trail, in which case both must have an odd number of edges. This is number is called the **degree** of the vertex. In our case, all the vertices are odd. Every edge we add, may change 2 vertices to be even. So we satisfy the necessary condition after 7 additions - changing 7 pairs and leaving 2 odd vertices. In our case, it is also sufficient by observing the following figure (added edges are dashed and the trail is illustrated by following the numbers).

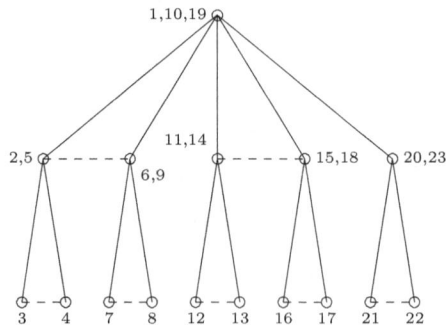

63). For a set $A \subset \mathbb{R}$, let A^o be the interior of A, A' the limit points of A, ∂A the boundary points of A, and $\bar{A}$ the closure of A. Which of the following is true?

I If A is closed, then $\partial A \subset A$.

II $\bar{A} = A^o \cup A'$

III If A has no limit points, then $\bar{A} = \partial A$.

(A) I

(B) II

(C) I, III

(D) II, III

(E) I, II and III

Answer is C. Observe the following Venn diagram that categorizes all the points composing $\bar{A}$:

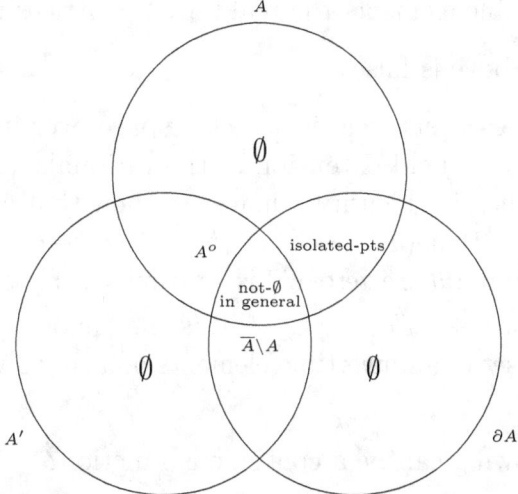

For I, the only portion of ∂A not in A is $\bar{A} \setminus A$, so if A is closed, $\partial A \subset A$. For II, we can construct a counter example by using isolated points. Take $A = \{1, 2\}$. A is closed, but neither of its points are internal, and neither are a limit point.

For III, Observe that without limit points, A is composed only by isolated points.

64). How many abelian groups of order 1200 are not cyclic (up to isomorphism)?

(A) 1

(B) 2

(C) 5

(D) 9

(E) 10

Answer is D. The devisors of 1200 are $2^4, 3, 5^2$. Using the known method for classifying abelian groups: For 2 - the number of partitions of 4 is 5. For 3 - the number of partitions of 1 is 1. For 5 - the number of partitions of 2 is 2. So we have a total of 10 abelian groups of order 1200. Only one of those C_{1200} is cyclic, so the answer is 9. Note that, up to an isomorphism, there is only one cyclic group of a specific order.

65). Which of the following is FALSE?

(A) $\mathbb{R}$ is a vector space over $\mathbb{Q}$.

(B) $\{a + b\sqrt{7} \,|\, a, b \in \mathbb{Q}\}$ is a field.

(C) The set $\{0, 1, 2, 3, 4\}$ with addition modulo 5 and multiplication modulo 5 is a ring.

(D) All the invertible elements the real 2 by 2 matrices is a group.

(E) None of the above is false.

Answer is E. For A - every field is a vector space over its subfield. For B - this is a classic example of a field extension of the rationals. The set is a commutative ring with a multiplicative identity. Suffices to show that every non zero element has an inverse. Observe that $(a + b\sqrt{7})(a - b\sqrt{7}) = a^2 - b^2 \cdot 7$ and the latter is zero if and only if both a and b are zero($\sqrt{7}$ is irrational). Thus $\frac{a}{a^2 - b^2 \cdot 7} + \frac{-b}{a^2 - b^2 \cdot 7}\sqrt{7}$ is the required inverse for $a + b\sqrt{7}$. For C - this is the famous $\mathbb{Z}/5\mathbb{Z} = \mathbb{F}_5$ field, let alone a ring. For D - The set of all invertible elements of a ring always from a multiplicative group.

66). Which of the following can be a continuous function?

(A) Bounded and surjective (onto) function $f : [0, 1] \to \mathbb{R}$

(B) unbounded function $f : [0, 1] \to \mathbb{R}$

(C) Surjective(onto) function $f : [0, 1] \to (0, 1)$

(D) One-to-one function $f : [0, 1] \to (0, 1)$

(E) None of the above.

Answer is D. For example $f(x) = \frac{x+1}{3}$ is a continuous one-to-one function as required. We can eliminate A since surjectivity on $\mathbb{R}$ implies that f must be unbounded. Both B and C are eliminated by recalling that an image of a compact set under a continuous function must be compact (in $\mathbb{R}$, a set is compact if and only if it is closed and bounded).

www.ingramcontent.com/pod-product-compliance
Lightning Source LLC
Chambersburg PA
CBHW080549220526
45466CB00010B/3084